PRAISE FOR RUNNING WATER

"Running Water is an amazing look at Abe Clark's inspirational journey. At times it seemed the encounters and events Abe faced were a fabricated story, but this was no fiction story. I could not put it down, in fact the only time I did was to go train. Finished it in one day. Abe's odyssey across the country sheds light on all kinds of cultures across America and equally shines light on a great cause for an important and necessary primal need, Living Water International. I would recommend this book to anyone, readers and non-readers alike."

Chris Solinsky – 8x Wisconsin high school state champion, 5x NCAA Division 1 champion, first non-African to break the 27-minute barrier in the 10,000 m setting the American record at 26:59.60

"No one could truly know what it is like to accomplish something like Abe did unless they've done it themselves. Although I have run across America myself, I had a support team with me who kept me under their watchful eye. Abe did his run solo. The two efforts are like night and day. I couldn't imagine doing what he did...wow. Simply amazing and for such an important and worthwhile, global cause. I am in awe of his accomplishment. He is a true inspiration."

Dave McGillivray – Race Director, BAA Boston Marathon, Run Across America – 1978

"Abe Clark tells it like it is in this compelling memoir of a solo run across the country. Clark relates each day, serving up equal doses of mundane reality and high adventure, just as it happened, mixed with the insights and self revelations that such an adventure reveals."

Running Times - Jonathan Beverly, Editor in Chief

"I am left with a big WOW after reading about Abe's journey across America. Being an endurance athlete who runs for a purpose I am inspired and enlightened by Abe. When you put a driven man such as Abe at the front on running for an amazing cause and purpose you can't wait to turn to the next page to see what lies ahead. Running solo is something that few people will ever attempt. "Running Water" will leave you wanting to get up and get going, I know I am going to!"

Lisa Smith-Batchen - 2x Badwater winner, Only American to win the 150 mile MDS across Sahara Desert, Only person to ever run 50 miles In all 50 states in 62 days.

"This book reads like a James Patterson novel; a page-turner. Not only do you get the feel of running, you also experience the variables of culture in America, the quirks of people along the way and the mistrust of someone running alone in the night. The journey, for a cause, turned out to be an experience of a lifetime, with ups and downs and a touching romance. Something we can all experience, if we read the book.."

Roy Pirrung - Ultramarathon world and national champion and record holder, with 60 national titles and over 60 American records and a member of USA Track & Field Masters Hall of Fame. 2011 runner-up in Italy's 722K (448 miles) Turin to Rome No Stop without crew support, at age 63.

RUNNING
WATER

a thirst quenching 2,960 mile solo run
across America

Adventure Memoir
Abraham Louis Clark

DEDICATION

This book is dedicated to those who have helped shape my desire to run. Coach Greiten and Coach Kline for helping me realize there is more to running than being first to the finish line and taught me that life is about the journey not the destination. Mom and Dad challenging me physically by paying me ten cents a mile to train when I was in fifth grade and for supporting all my adventures. By the way you owe me $15,000. Ben and Josh for years of brotherly competition and high standards. Rachel and Sarah for being built in cheerleaders. Kate for believing in us, my dreams and our future. Jason Soper, boy was it an exciting six years of racing, thanks for all the long training runs. To the people who helped me along the way, prayed for safety, who called ahead to find housing and the ones who offered randomly to help me and finally the ones who donated to Living Water International. Lastly, this book is dedicated to the 884,000,000 people who do not have access to clean drinking water and the Living Water International staff who works everyday to make that number shrink in Jesus' name.

ACKNOWLEDGMENTS

Editors – Margie Clark, Jennifer Boettcher, Rebecca Rousseau

Photographs - Jon Carpenter, Josh Clark, Abraham Clark

Product Sponsors - Reliv, PowerBar, Sugoi, Serfas, Fuelbelt, GU, Songear (Bob Van Zandt), Bodybasix, Go Healthy (Eric Mader), Juice Plus (Joe Harper)

www.stridersrun.com - Thank You Steve Webster and Striders for providing me with 2,860 miles of Asic Cumulus running shoes, socks, energy products, running shorts, jacket, skin care products and more!

www.babyjogger.com (best stroller EVER!)

CONTENTS

PROLOGUE

Attending a small three room rural grade school in Northeastern Wisconsin as a child gave me a unique perspective of life. With a population of 605 the small town had a knack for breeding dominating basketball teams in our conference; however, the hard wood floors and brick walls of the classic small town gym never lit up my imagination so during my 5th and 6th grade years after school, my mother would drive me ten miles to the next town to run with the freshly established Gillett High School Cross Country team which my older brother, Ben, had joined.

Bill Greiten, a Wisconsin Hall of Fame Coach, had moved to the northwoods to retire. However, he could not resist starting a Division III program in our community, a community that knew little of distance running. During the 70's and 80's his teams dominated the Wisconsin's distance scene. They grew long hair and piled up state titles like clockwork. He demanded excellence out of his rough bunch of athletes from the inner city and always pulled more out of an individual than they were usually willing to give. He was a legend in the scene of Wisconsin cross-country. When our paths first crossed, it was on the old railroad bed that passed by the school and continued 32 miles north, passing my house along the way.

My young mind was ready to be shaped. I was years younger than the high school runners and was usually the only one who would not be able to out run the elderly coach on that endless trail. So there we were, an odd running couple from the looks of it, he having way too many miles on his legs and I too few. Stride for stride, every day after school, this larger than life man would fill my head with years of running stories. He brought out the desire in me to run and instilled in me the knowledge of what it took to go faster and farther than anyone else. He told me once "A person only has so much God given talent, it's up to you to find how much that is." It sent me into life with the mission of searching for my own limits. The following account is the closest I have come to finding the illusive limit of my heart, body, and most of all, my own soul.

CHAPTER 1

ONE LAST LOOK WEST

July 29th 2009 - I reached over the side of my 26-foot sailboat and into the world's largest body of fresh water. The cool clear water darted around my fingertips as the wind pushed us over the sparking waves. The sandy beach of Chambers Island, where I had proposed to Kate the night before, faded in the distance. We were silhouettes caught in yet another sunset designed just for us. Kate looked down once more at the diamond on her ring finger and curled up close to my side. I shook my wet hand off in the late summer breeze and placed it around her shoulders. Pulling her in tight, our eyes met and locked for the hundredth time that day as she softly asked, "Do you think we will change the world?" She had brought up the question many times; it was her way of asking me what my deepest thoughts and wildest dreams were. I had been living on the boat for the past few months, sailing it from port to port and island to island, watching the distant coast slowly slide by. During this time my thoughts often sought out the answer to that question. I repositioned my foot on the tiller.

Seven Months Later - It was Monday morning; I stared out over the Pacific Ocean, slightly squinting my eyes from the early morning sun that shone through the cracks of the longest wooden pier on the west coast. My bare feet slowly sunk in the cool mid-February sand, the continuous waves washing around them. Perhaps the sand would bury me right then and there if I lingered long enough, thus saving me from turning and facing the continent that I pledged to run across. I leaned over and filled a small bottle with a piece of the ocean. It would be a constant reminder for me of how far I had come in the days, weeks, and months ahead.

Day 0
"This is the beginning of a journey that will test my limits, challenge my body, play tricks with my mind and match no other extreme physical experience I have ever had. By the time it's all over, I will have literally climbed the amount of elevation equal to reaching the edge of outer space. And become only the 15th person to cross our nation in this way. The opportunity is incredible, but the odds against me are great. This is the position I wanted to be in, now is my time to live!"

A Ruby's restaurant at the end of the 1,942 foot wooden pier inspired the name of the companion that would accompany me on the 2,960 mile run. She was a red jogging stroller which carried all the essential gear I could get my hands on. With a glance you could tell that this was no high scale operation. She was no team of doctors or cooks ready to serve at the snap of a finger, but she was my own version of a home on wheels and I was thankful to have her along.

I had been staying with my older brother and his growing family during the days leading up to the run. The final year of his commitment to the United States Marine Corps was finally here. Loving the outdoors as he did, San Diego California was probably the last place on earth I ever thought he would live. My theory appeared to be substantiated by his well thought out plan to escape back to the north woods to build a cabin as soon as he was discharged.

Jon was also along on that first day, a friend from back home, who enjoyed photography along with an array of various outdoor sports. He had flown from Wisconsin to document my send off and capture my innocent and eager state. My pale Wisconsin winter skin and a smooth shaven chin were a few of the more visible aspects that time and miles would eventually change.

I dried off my feet and put on my first pair of size 13 Asics Cumulus running shoes. They would hopefully carry me across California and into the desert. I was eager to get going and finally put those first few miles behind me, but before I crossed the road, an elderly man called out to me.

He introduced himself as Bill Ash, which didn't mean much to me at the time. He seemed to be extremely interested and knowledgeable about Living Water International, the charity I had chosen to help raise awareness and money for during my run. He gave me a picture of a small African boy on his hands and knees drinking from a puddle of muddy water. We chatted for a few minutes; then I took off down the road.

Days later I found out who Bill Ash really was- he had founded Life Water in the late 1970's after taking his family on frequent outings to Baja, Mexico to help orphanages, camps, and churches install new water systems. The family's water pump business was prospering in Southern California; and Bill and his wife, Lorraine, wanted to teach their children that such blessings brought with it an opportunity to give. It always fascinates me how the things we choose to do in life can snowball, whether good or bad. There is only room in life for good or evil; and the more good we spread, the less room there is for evil. I am sure back in the 60's there were times when Bill sat in the basement of an orphanage somewhere feeling defeated by a pipe that did not fit or water pump that just would not work.

Little did he know those efforts would lead to thousands of missionary volunteers, multiple water relief missions, and millions of dollars donated in the name of Jesus. Life had taught him that it was the little things that mattered. It had taught him that although walking was now difficult, driving two hours through L.A. traffic in hopes of giving some skinny runner a photo he took years ago was well worth the effort. As it turned out, that photo gave me strength in some of the darkest times of my run. It is the image of a single child that will be forever imbedded in my mind representing a group of people so large that every step I took towards the Atlantic Ocean would represent 200 people who lacked access to clean drinking water. Bill Ash was a wise man.

A block into the run, an orange glowing hand on the streetlight halted my adrenalin rush. As the polished California cars zoomed past, I could not help but doubt my preparation. My training runs leading up to the run had been spent slogging through snow often over my ankles. Heavy with winter gear, I would slowly pick my way down a single-track trail that ran along the Milwaukee River. The trail was 20 miles of wooded escape ribboned under the concrete bridges, tall square buildings and street sounds of the city. Although I was in the best shape of my life, today's 30 plus mile jaunt would be my farthest run ever. The light changed, and I jogged across the intersection before clumsily smashing Ruby into the curb. All of my sponsors came on board so late that most of the gear, including the stroller, had to be shipped straight to my brother's house. I had never run with a jogging stroller before; it felt awkward at best. The fixed front wheel needed to be perfectly aligned in order for it to roll straight. Small rocks or cracks would throw it off track causing me to constantly correct its direction. This was easily done by pressing down on the black rubber handlebars and periodically bouncing the fixed front wheel back on track. The task seemed to require too much focus at the moment, but would soon become second nature as Ruby herself almost became an extension of my right arm.

Stoplights became sparse as the beach culture turned into country homes surrounded by orange orchards. The hills rose and sunk as if mimicking the fading waves of the ocean.

Day 1 – 33 miles

"I was on the road for eight hours today. Jon picked me up and we drove back to Ben's house in the dark. Sarah and Ben had a feast of chicken waiting for us! I treated myself to a bathtub full of ice water and almost fell asleep in it. Today's 33 miles was the farthest I have ever run in one day. It's a great feeling to be on the road and put the months of questions, anticipation and preparation behind me. Well, it's 9 pm and I'm off to bed. Tomorrow we will do it all again. THIS IS RUN CAMPING!"

In the morning I said good-bye to Ben, his wife Sarah and newborn son, Caleb. Ben being the oldest of my four siblings has always played a protective and somewhat cautious role. Of every one I knew, he discouraged the run the most. He said he just didn't want to drive five hundred miles into the desert and pick up my half dead body, but I knew it went a little deeper than that. Jon and my Aunt Kiki dropped me off in the morning, and I began my first day completely on my own.

I ran along the bottom of a few populated valley's separated by hilly ridges. The ridges, long and gradual climbs, proved to be easier to run than the steep hills that congregated near the coast. With my second day on the road closing to another California sunset, I was dehydrated, sun burnt and out of energy. Sometimes in life things seem to come along at just the right time and as I walked into the air-conditioned gas station 60 miles from the cool ocean breeze, God answered my prayers. There she was, love at first sight. A giant Big Gulp, other wise known as shredded caffeinated blue raspberry ice! Just like that my mind was free of dragging miles and my body revived as blue raspberry ice shot through my swelling veins. I was back! There are basically two speeds to get where you are going, slow or fast. As the bright sunny day ended and turned into a consuming black night sky, I was going fast!

Day 2 – 30 miles

"I waited on the corner for my host family to pick me up. Chilto, my Aunt Kiki's friend, from the Peace Corps, arrived around 7:30 and ordered me a large pizza, which I ate in its entirety. They offered me a shower and a large double bed. I have no complaints."

Through the first few days there were certain things that gave away my lack of experience traveling by foot. Chilto and I dropped the children off at school and headed out to where she had picked me up the night before. When I took out Ruby and began to pack her up something was terribly wrong. I had forgotten Ruby's detachable 20 inch wheels in the garage and she sat there limp as a dog sled without snow. I had no choice but to ask if we could drive all the way back to get them.

My late start was again interrupted when I came to a locked gate blocking my entrance to Lake Perris State Recreation. I was hesitant to crawl under the gate and proceed through the park but was left with few options. A ticket window read the list of fees for the park, $35 for camping, $20 for boating. I decided if the DNR did stop me, I would just have to kindly let him know that there was no fee posted for travelers on foot.

Lake Perris looked like an oasis in the desert. For a Wisconsin boy, the weather was great for camping and summer activities; however, for everyone in California, it was winter and a frigid 70 degrees. With the locked gate behind me I had the whole road to myself. This was the first time in my three-day 86 mile run that there were not people around. I smiled to myself; I was completely free.

Eventually, I wound my way out of the park, through Moreno Valley and over the Live Oak Canyon pass. The views were incredible, the San Bernardino Mountains were a backdrop to hot air balloons drifting over the valley. Ruby and I rolled into Yucapia as the sun dropped behind the towering mountain peaks. It caused the entire town to be engulfed in its shadow and was a harsh reminder of the road ahead.

Doug found me asleep on a small grassy strip that ran along side the church where we were to meet. We loaded up Ruby into the truck and he drove me to his home. Homemade turkey soup hit the spot and I was once again grateful for a place to stay for the night. I had planned to go over the San Bernardino Mountains; however, Doug was very determined about changing my planned route. A snowstorm had just dumped six feet of snow on the mountaintops closing many of the roads, including the one I was planning to run on. The trade off was extra mileage and possible freeway stretches.

Day 3 – 23 miles
"For the amount of mileage I put in the past 3 days, I feel pretty good. I have some minor swelling in my feet, sunburn on the back of my ankles and a very sore back. My wrists feel hyper extended from pushing Ruby. I'm ready to hit the hay."

The decision was made. I would attempt to go around the

mountains. For the most part it was a gradual downhill route excluding one 3,500 foot climb. It was another perfect sunny California day. I was excited to know I had a host family in the next town, a luxury that I would not have had if I had gone through the mountains.

Everything was going smoothly until four miles to go. My luck of running on service roads along the freeway had come to a literal dead end. I had expected this from looking at the satellite map the night before with Doug but didn't expect the barbed wire fences. There was no clear-cut road besides the freeway. A mess of dead ends and dirt paths lead nowhere. I found myself pushing Ruby through a huge field only to find at the other end an insurmountable fence. Fat Ruby made the six-foot high barbwire into the Berlin wall. I found a dry run off ditch and after removing Ruby's back wheels, I was able to squeeze through. The rough terrain continued until I stumbled across another service road that lead to the A&W that Doug's friend, Kevin Allen, was to pick me up at.

Day 4 – 22 miles
"It was awesome to have so much of the route be a slight downhill today. At times I was doing 7-minute miles, until my road ended. I find humor in the reactions of people driving by. Some smile, some wave, some shake their fists, some give the peace sign and some simply point. One thing they all do is hold their stare a few seconds longer than socially acceptable! What can you expect?"

Mountains climbed high on either side of the valley causing a wind tunnel that would have blown a sailboat head over heels. My route was the prevailing issue again. Dead end roads and barbwire fences quickly became much more than a pet peeve. With a determined rush of adrenaline, I would throw Ruby over a fence and continue down the dirt road that serviced the billboards along the freeway. This type of travel went on all morning. I was fighting for every foot. None of these roads were on my GPS and the whole day proved to be completely frustrating.

The wind was ripping through the valley, and I could hardly hear myself think. I was trapped in an endless wind tunnel of speeding semi trucks and giant windmills. When I thought that it couldn't get any worse, I came to a fork in the road. On my left was a steep single lane dirt road leading up to fields of windmills. On my right was a sign that read, "Road ends in 500 ft."

I started up the dirt road when a huge dump truck came around the turn a quarter mile ahead. There was no shoulder on this road. I instantaneously turned around and bolted down the hill. My second option turned into the ditch of the freeway. Then the ditch disappeared being replaced by the foot of a mountain. It was an impassable route. Ruby admitted she was no match for 80 mph semi trucks and I certainly was not fond of the idea either; however, I was out of options. I tightly hugged the edge of the freeway and took off. I ran fast, easily completing 5-minute miles. The wind was howling and cars were blowing by. I stared straight ahead whimpering, "This little light of mine, I'm gonna let it shine." The mountain finally gave way to a ditch and I thrust Ruby into it. The ditches were extremely rough and it was a true test of how durable Ruby really was. I had finally made it to old 29 Palms Hwy. The road was empty, the scenery was breathtaking, and I reached for my camera... It was missing.

I couldn't believe it! I had just come through the worst stretch imaginable and had to go back. My heart began to race as I wheeled Ruby into the bushes. It was a risky decision to leave her, but I had to go alone. I ran back over the way I had come, my eyes scanning the desert ground. I wound my way through the mess and continued searching. After two miles I put my hands on my face and slowly began to lose it. I was looking forward to documenting the trip with my camera.

I looked up to scan the area one final time and astounded, saw my camera a few feet ahead under a bush. A big sigh of relief and I turned around to catch up to Ruby. I knew I had one more climb before Morongo Valley and the four miles of highway that cut through another ridge of hills would test my limits. Traffic was fairly heavy and it took a lot of will power to finish off the day.

The last of my energy dropped me off in the back of Monongo Valley's wildlife viewing park. I wandered around for three hours in that treeless park trying to find a place to string my hammock. It would be my first night camping, and I was excited to try the Hennessy tent hammock my younger brother, Josh, had lent me. After changing my hammock location four times, I finally settled for a less than perfect location. My lack of efficiency magnified my inexperience. Yes, I was from the north woods of Wisconsin where finding a good hammock location was effortless; however, was that any excuse for thinking there would be trees to hang it on in the desert? Nevertheless, a hammock was what I had for shelter, so I did my best to make it work. Two huge hills that came together formed a slopping ravine. I tied one end to a three-foot palm tree, crossed the ravine, and tied the other end to the base of a bush.

It was almost sunset as I unloaded the necessary gear from Ruby. A couple handfuls of trail mix and a few pieces of beef jerky would serve as my meal for the evening. I knew it would get cold so I put on all the clothes I had. I also took my 10-inch flat bar knife, bear pepper spray and crawled into my shelter for the night. It was outlandish trying to get my sleeping pad and sleeping bag properly arranged in the swinging tent-hammock. I was exhausted and out cold the minute I was situated.

I woke up to find myself barely off the ground with little assurance that I wouldn't end up rolling down the ravine. I checked my watch and couldn't believe it; it was 8 pm. I was in for a long night. The wind would blow up under my rain fly and lift the thin layer of insulation exposing me to the cool night air. If I breathed into my sleeping bag and didn't venture any body parts off my sleeping mat, I was warm. The night wore on. At one point I heard cats meowing but thought nothing of it until the next morning. As I walked out of the park a sign read, "Mountain lion warning, do not venture from trail." I was happy I hadn't discovered the sign yesterday. Wrapping one's self into a swinging cocoon of fresh meat for mountain lions to gnaw on is probably not mentioned in Bradford Angier's book, *How to Stay Alive in the Woods*. I found my way out of the park, with only a few stares from avid bird watchers, and was back on Hwy 62. I felt accomplished having survived the night.

I jogged out of Morango Valley, and a green sign read 32 miles Twenty-Nine Palms. The highway was wide and long with gradual sweeping hills. It was all runable and I made good time. My hips ached from sleeping pinned between the hill and the hammock. Fortunately, my brother found two Marines for me to stay with in Twenty-Nine Palms, and I couldn't wait.

Day 7 – 136 mile
"I stayed with a Marine family all of Sunday and lay around like a lazy bum. I can't believe I'm in Twenty-Nine Palms, it has been quite the week! Over the course of 6 days I ran 136 miles from ocean beaches, over hills, through valleys, around mountains and was now entering the desert. It feels like I have already crossed a whole continent and been on the road for a year. There is no way to tell how I will feel months down the road. The Marine families feed me lots of pizza and made me feel right at home. I must say our country is in good hands! I would never want to fight someone who uses a .50 cal shell as a staple remover. Chris took me to the store to stock up on PowerBars, fruit, and an extra

jug of water for the desert. I was rested and ready for week 2! It will be a long haul to Bullhead City, and it lookslike I'll be camping every night."

Above - Hiding from the wind under a freeway bridge in California. Creative graffiti of a giant rat painted on the wall behind me.

Below - My first sight of a road that appeared to have no end, many more would follow.

CHAPTER 2
NEXT SERVICE 90 MILES

On the road running at 8:00 am, my legs were feeling pretty good. I think the day off really helped. I was headed out of Twenty-nine Palms to Amboy Road. Even on the map Amboy Road looked long and desolate. The first sign I read was, "Next services 90 miles." Stocking up on provisions and water in Twenty-nine Palms was a good move. Cars were scarce, and I was able to enjoy the ideal running road, putting 20 miles in before noon. However, the miles were a bit harder to come by in the afternoon. I had made the only turn on Amboy and was now headed north. The wind picked up, and I was not as fortunate to have it at my back as it had been in the wind tunnel of Interstate Ten. The wind may have been bearable but for the fine mixture of sand in it. The funniest things would get caught in the wind and pass by me traveling in the opposite direction, one of them being a car bumper. I'm not sure if it blew right off the car or if it had come from some far off junk yard. I crossed through the salt flats and over a huge hill. After hitting the peak and beginning my descent, Ruby still needed to be pushed down the hill due to the strong head wind. The landscape was simply huge! I was entering what looked like an empty Lake Michigan. About halfway across I decided to set up camp and get out of the wind. My previous night in the wild was a crash course on what not to do, and I was excited to put my freshly acquired knowledge to use.

It was still pretty cold at night, and an extra layer would have been a luxury. I had five t-shirts and wore every last one. Since there were no trees around again, I decided it was better to be completely on the ground rather than attempt another insufficient set up that would inevitably leave my body awkwardly hovering half on the ground, half in the air. The only vertical structure in sight was one of the telephone poles that supported the endless wire. Tying one end to the pole and the other to Ruby's wheel I turned the hammock into a small lean to or tent. Dreaming of the conversations that whizzed by above, the hum of the power lines put me right to sleep.

Peeking out of my cocoon at sunrise I found that the wind had died down to a calm breeze. The night went a lot better compared to my evening at the wildlife refuge. It was 7 a.m. and I hit the road. Telephone poles turned into toothpicks as they disappeared over the horizon. It took me a long time to get loosened up, and I only made twelve miles before noon.

Amboy Road eventually dead ended and left me at Roy's Gas Station on Route 66 in the unincorporated town of Amboy. It was one of the survivors from the glory years. Outside the single pump station a large sign read "Café". I was excited to fill the void in my stomach with a sandwich and sit out the mid day sun. Much to my disappointment, the only thing filled was my ear. The lady working at the place was a talker and filled me in on why the café had been closed for some 40 plus years. Too tired to go anywhere I listened to the whole history of the place, from its 1926 boom town roots through its peak population of 65 and right up to her. She would drive 75 miles from Yucca Valley to work at the gas station for a couple of days straight while staying in a trailer out back. She described it as a labor of love keeping the history alive and telling the story to anyone that would listen. We took a picture in front by the sign for their Facebook Fan Page. After I mailed a post card from the other building in Amboy, the post office, I finished my soda pop and orange and took off down Route 66.

I did not know what it was about the road but I was excited to be on it. Maybe it was all the strange items that peaked my interest along the route. There were only a few trees but when I spotted one it, usually was covered in strange things like underwear or shoes. People would write their names on a mound that ran along the side of the road. They would use anything from colored rocks to clay pigeons to make their name stand out from the pile of dirt. I thought it was silly at first but after a couple of miles I started reading every single name written. I would wonder who they were and what they were doing now.

Day 9 – 34 miles

"The roads are long and straighter than an arrow. Today I ran 34 miles and only made two turns. My body goes through these strange energy cycles. One minute I am rattling off 5k's like I had been training for a race. Then after about 25 minutes of this, I slow to a painstaking jog. Eventually I walk and sit down on the side of the road and take a picture or have a PowerBar or gel. Then 15 minutes later it starts all over again. I did this all day long."

The sun was getting low in the sky as I crossed a ridge and stepped into a vast bowl that looked exactly like the one that I just spent a full day crossing. I ran three more miles shirtless and decided to set up camp under a bridge. It had a roof, two walls, and strong poles to hang my hammock on. I'm not sure why but I felt safe being connected to the road. Maybe I was spending so much time on it during the day my brain just assumed I should be under it at night. I built a giant fire, watched a distant train roll across the desert night and stared up at the biggest ring around the moon I had ever seen. Folklore has it that a ring around the moon signifies bad weather is coming; I guess it is only a matter of time.

I awoke to the sound of semi-trucks rumbling over the bridge at 4 a.m. Slightly bored and a bit cold, I crawled out of the hammock to rebuild the fire. After watching the sunrise I packed up Ruby and hit the road. The sleeping location worked out well, and if the opportunity presents itself, I'll try the bridge method again. I can see why it has become a popular shelter for the homeless.

It has been cold in the mornings, and I usually leave night gear on for the first few miles before slowly shedding layers as the day goes on. I occupy my time by playing games like trying to keep Ruby's front wheel on the white line.

I had never been more excited to make it to a town where the population sign read 100. I think they rounded up. I walked to the only place open, JB Tire. The guy built the place himself in 1961, and I'm sure at some point the business had its glory years. Our dialogue went something like this, "Do you have an outlet I can plug my phone into?"

Sitting in an old dusty chair he replied, "Right over there in the wall."

My response, "Thanks."

He bluntly fired back, "Don't park your buggy there, I might get a customer." Now the other ten citizens of Essex may have been loyal customers of the old tire garage but I had my doubts that there was a customer within a hundred miles of him.

The next ten miles felt like the earth had doubled its gravitational pull, my legs were heavier than logs. I kept thinking there were Ninjas hiding on the side of the road throwing knives into my legs. I don't know where these shooting pains were coming from, but they had no rhyme or reason and usually left as quickly as they came. However, I had housing in Goffs and was determined not to spend another lonely night in the desert. When I got to interstate 40 there was a slightly normal gas station where I sat in the shade and stared out over the Mojave Desert. An hour passed maybe two before I left. I bought a hot dog and Snickers bar and it made for a good lunch when combined with a PowerBar and energy gel. A few minutes later I was flying down the road to Goffs and throwing my pumping fists in the air to get a horn blast from the passing by train conductor. I'm not sure what they put in those energy gels, but it was awfully powerful for three or four miles. I could see Goffs in the distance and was thrilled when two older men came out to meet me in their golf carts.

Day 10 – 29 miles

"Tonight I am staying at a historical museum park located here in Goffs, California. It is an outdoor museum with all sorts of old things ranging from the pioneer days through the rise and fall of Route 66. Hugh Brown gave me a private tour. I learned all sorts of interesting things like how they used to get gold out of rocks and how the city of Goffs came to be. Hugh showed me to my own personal 16-foot trailer where I would stay for the night. It reminds me of living on my sailboat last summer with its miniature cupboards, sinks and bathroom appliances. After a quick outdoor shower I headed over to the red boxcar for dinner. There were five of us and we had quite the feast, sparkling cider and all. We talked for a long time about the good old days and had some pretty great gut busting laughs. The people out here seem hard up for entertainment yet strangely okay with its absence. It's really quite beautiful."

It was probably a mistake to use my bag of energy bars as a pillow because I ended up eating four PowerBars during the night. Taking advantage of the great set up, I didn't hit the road until 11:00 a.m. The first 15 miles of the day were dedicated to finishing off Goff's Road in two and a half hours. Turning left on to US 94 to head north, I spotted what would most likely be the only place to string my tent hammock. It was an old cattle pen about a quarter of a mile off the highway.

Thoughts drifted to Hugh Brown's tour, and I wondered if the place was left over from the Homestead Act of 1862. The two tallest posts that made up the decaying sun dried fence appeared to be the entrance to the pen at one point. For now it would serve as my hammock support. Keeping warm at night was a challenge. By natural instinct I was quickly learning tricks of survival. Wrapping the emergency blanket around the hammock before the rain fly went on made a deafening sound in that empty Mojave Desert but cut the wind and added much needed insulation.

Wood was plentiful because the remains of dried out fence pieces littered the area. In no time at all a blazing fire sprung into life. I hovered close; my face and jeans feeling the intense heat, leaving my jealous back to greet the unwelcome cold desert air. Water began to dance in the small tin pot that was jammed full of spaghetti noodles. It was the first time I opened the box of noodles since buying them in Oceanside eleven days ago. If you had asked me if it was worth carrying those noodles some 300 miles I would have shouted, "YES! Yes, it was." The warm noodles slid down my throat and landed with a thud in the bottom of an empty stomach. Stars and endless black dissolved through another orange desert sunset. A peaceful gaze at the scene unfolding was interrupted by the howl of a coyote a hundred yards off. Caught off guard I reached for my knife and pepper spray while tossing a few more sticks on to the fire with my other hand. They were harmless; nevertheless, I didn't want to wake up to anything in my hammock with me.

It was a harsh reminder of where I was.

The sleep eating PowerBar incident the night before left me with no food for breakfast. It was cold out, and I ran the first two miles with my jeans on. It was getting harder and harder to convince my body to run. Eventually my legs would regain their momentum and remember the forward motion they were starting to know all too well. The sharp pains continued from time to time. A painful shot to my high inner thigh brought me to the ground screaming. My voice went unheard, lost in the white noise of passing semi trucks and oversized motor homes. Time passed as I sat on the gravel shoulder, analyzing my pathetic state. The desert of eastern California had taken its toll on me.

Now only a half-day run away from the boarder of Nevada, my body and mind hated what they were forcing each other to go through. Beginning to lose my mental edge, food and a normal place to sleep was all that was on my mind. Both of them were about 20 miles down the road in Laughlin. Craving anything that resembled food I ate what I had, energy gel. Consuming seven in a row I took off my jeans, turned off my brain and ran. Jogging into Laughlin, Nevada at high noon, I walked into the first restaurant with an open sign. It was a Mexican place with a large picture in the window of a double beef burrito. Appearing to be the largest thing on the menu I ordered two. Honestly the burrito tasted awful, but my body welcomed the food. Energy literally rushed through my veins and filled the empty shell that I once knew as my own body. Next door for 38 dollars they promised me a room with no wild animals. The receptionist couldn't seem to grasp why I didn't have a license plate number to give her. I didn't care to explain. On the wall was a thermostat that was capable of making the temperature in the room anything I wanted. I crawled into the over-sized bit of heaven called a king size bed, and ordered a large take out pizza. Before my eyes closed the last bite slid down my throat.

Day 12 – 24 miles

"Laughlin is a crazy town! People seem to come from all over the place to test their luck in the big casinos that overlook the Colorado River. There are signs everywhere saying things like, "You could win $250,000 with the swipe of your card!" Picking out who actually ended up winning is as simple as looking at their transportation. One tourist told me she had come a long way to participate in the big poker tournament that would be held that night. She wore her fanny pack low, mainly because it did not fit around her mid section. After wishing the enthusiastic character good luck I couldn't help but smile to myself as she drove off in a run down motor home with duct tape on the fender."

After a quick morning stop at Subway I took my two foot long subs and made my way towards the bridge that crossed the Colorado River. The river was crystal clear, and I would have gone for a swim if the weather had been anything close to the normal Arizona heat I had planned on. Instead it was a cold and rainy day. The rain came down in sheets to the south.

In an attempt to stick to the smaller roads I decided to take an alternate route through the Black Mountains. The road turned out to be unpaved and wound through the mountains like a slinky. Ruby would not roll an inch by herself leaving it up to me to pick up her slack.

The hills grew as I approached; a man shooting a handgun at an unknown target beside the backcountry road took time to stare me down. Anyone shooting a handgun in the rain must have been some kind of loony.

He was probably thinking a similar thought about me. His gunshots eventually faded due to my continuous forward motion. The GPS said I had come 17 miles but when I checked to see the distance to the next town, Kingman, my jaw dropped.

"31 miles? That can't be right! When I left the hotel it was 32 miles." I yelled out loud to myself. Something had gone wrong. I was steaming mad and took off down the windy, gravel road at a reckless pace. Entering into a desperate mindset I told myself I would not stop until I got to Kingman.

At sunset I stopped to change my shoes. Reaching under Ruby's rain fly I pulled a new shoe out and reached for the second. It was gone. My right shoe must have bumped out of Ruby unnoticed somewhere along the eight miles of gravel road that I had just run over. Not thinking clearly, I left Ruby on the side of the road and turned into the setting sun with nothing but an energy gel. A mile and a half later I found nothing. A few cars had passed by making me realize the mistake of leaving Ruby unwatched. A fear came over me that someone would steal Ruby or the things she carried. Everything I had was necessary for my survival. As I came over the last hill I saw four people standing around Ruby.

They started waving and shouting my name "Abe! Abe!" As I approached they all started talking to me at once "We thought you were dead down in the canyon or something!"

"We couldn't figure it out."

"We called home on your cell phone."

After speaking very little with anyone for days on end, the unexpected joy and excitement they brought as the sky grew dark was overwhelming. Doing my best to explain my situation, I described how I had lost my much-needed running shoe. Erik Laybourne offered to take his heavy-duty pick up truck back over the gravel road and search for the shoe with his friends. They all jumped into the truck and sped off down the hill and out of sight.

Just like that, I was alone again in the hills, the silence now much more noticeable in my ears. Not knowing if I would ever see the energetic group again I turned to the road not yet traveled and jogged into the now darkening hills. Before I could begin my long reoccurring process of self-pity, headlights shot up over the hill. Recognizing the roar of Erik's oversized engine I knew who it was.

The truck came to a halt, and they yelled out the window, "We found it! We found your shoe!" Sure enough there it was my lost orange Asic Cumulus running shoe. Erik smiled and asked, "Where are you staying tonight man?"

"I'm going to try to make it to Kingman," I replied. His smile turned to a laugh followed by serious concern when he realized I wasn't joking.

"There's no way you are going to make it through the Black Mountains in the dark… It's over 30 miles!"

I looked down the road and said, "I know I just don't feel like camping in the cold tonight, I'll sleep during the day." He opened his silver door and jumped out "We have a summer house a few miles from here, would you like to stay in our guest room?"

My eyes shifted from the road to his eyes "I would love that. But you have to bring me back here in the morning."

"No problem," he said as we threw Ruby into the back of his big truck.

Day 13 - 20 miles
"It was a crazy end to a crazy day. Thank goodness for the shoe angels!"

The turn of events the day before put me in the position to postpone my planned day off and make one more 28 mile push to Kingman. Erik told me all about the strange things in the Black Mountains that I would witness later in the day. Back in the day, the Black Mountains were full of gold. The miners would pack all their equipment on mules and head into the hills to begin their search. When the mines shut down, the miners left behind their donkeys and to this day there are wild donkeys that roam around Oatman and the Black Mountains. It was the funniest thing to watch those donkeys hopping around on the hills and listening to their hee haws echoing throughout the mountains.

Originally the road was part of Route 66 that cut through the Black Mountains. People coming through would pay the locals to drive their car over the mountain roads. Eventually Route 66 was changed to go around the Black Mountains on a much safer road. The stories didn't cross my mind again until I got up to the narrow switchbacks with blind corners. Around every turn there was a sharp drop off with three or four old cars left in a pile of crumpled metal far below. The images of the wrecks kept my guard up and I changed sides of the road frequently to avoid getting hit by an oncoming car. I passed a gold mine that was up and running. It used to be open for tourism but when gold got above $450 an ounce they decided it was more profitable to pick at the rocks than the tourists' pockets.

They were not the only ones looking for treasure in those mountains. About halfway down the other side I came across four men chipping at the rocks with chisels and hammers. Squeezing Ruby's hand brake the down hill run away stroller came to a stop next to the hole the men were in. They paid no attention to me, continuing to swing their five-pound hammers at their chisels. A large bellied man waddled out from the shade of an over hanging boulder. Unlike the men in the hole his sweat soaked shirt was a result of simply being alive rather than putting forth any manual labor.

"What can I do for you?" he said over the rhythmic pounding.

"Just passing by wondering what you guys were up to. Are you looking for gold?"

As if an answer would have taken far too much energy in the midday sun he simply held out his hand. On his finger he wore a large stone. Out of breath from walking around the hole he managed to proudly state "It's a semi-precious agate.

This is my hole, been digging in it for four years."

Struggling to look impressed I replied, "Well it's a beautiful rock sir, I sure hope you find more."

"I will," he said as he turned to head back to his shady chair. "Just letting you know this is my hole" I certainly did not have the least bit intention of claiming his hole as my own and decided to leave the awkward encounter and continue on.

The road stopped winding when I got out of the mountains and shot me across a vast open valley. I hate to say it but the 12 mile stretch to Kingman was a little boring compared to the Black Mountains. I checked my fatigued body into the Arizona Inn and did not plan on moving a muscle for two nights.

Hundreds of names line Route 66. Taking a break next to Ruby's

Running into the night with the sun setting at my back.

Entering the Painted Desert on the Navajo Reservation. The rocks were bright red, a sharp contrast to the black road.

CHAPTER 3
THE RUNS

Out my motel window I could see a local museum. Curiosity eventually over took my desire to rest and I rolled out of bed, slipped on my running shoes and crossed the street with shoelaces dangling.

Two elderly women volunteered their time to sit at the entrance of the local establishment and greet people as they came in. Whether they realized it or not they also gave me an audio tour of the current local gossip as I slowly strolled around the small museum. Most of the exhibits were about the Black Mountains and Route 66. The town of Oatman was named after the Oatman family. The Apaches kidnapped Olive Oatman and sold her to the Mohave Indians a year later. Over the next four years of her captivity they fed and clothed her but also tattooed four vertical blue lines down her chin. Later, at age nineteen, she was "rescued" and returned to white civilization. But with her tattoo, she was literally, for the rest of her life, a marked woman.

Being the only excitement in the place there was no way, I was getting out the door without telling my whole life story to the two women. One of them took $5 out of the register; handing it to me she said "Nobody leaves Kingman without trying one of Mr. D's burgers. We stop in every Monday,

Wednesday, and Friday." I had not eaten yet today and my
stomach nearly reached out and grabbed the five bucks before
my hand did. As I walked out, I wondered if they paid for all
their local gossip or were just extra fond of me. Either way it
was a kind gesture and brought a smile to my face.

Mr. D's was a 50's burger joint on Route 66 with bright
pink and turquoise colors. It was a shrine to Elvis Presley and
the cars that used to drive up and down that old road. The
burger wasn't anything special, but I have to admit the sweet
potato fries were the finest I have ever tasted.

Day 15
"It is nice to have some time to take a step back and
reflect on the first two weeks. I ran/crawled 355 miles
since taking those first couple steps out of the Pacific
Ocean. It already feels like the journey of a lifetime.
It still surprises me how far I can make it in one day.
I need a better method of packing Ruby. Things keep
falling out and if I don't notice it until later down the
road the decision to backtrack and search for it, reverse
or move on without the item is agonizing. I'm pleased
that my body has made it through the initial shock of
extreme mileage! From here on it's more of a mind
game, 20 to 30 miles a day is a long way no matter how
you look at it. I have to mentally face that challenge
every morning. Whether I feel great or lousy the same
long road is still before me."

It didn't take long for me to realize one of the biggest
differences between being solo verses having a crew with a
chase vehicle. I desperately needed to restock my provisions
before heading back into the empty desert of northern
Arizona. My only option was a Wal-Mart a couple of miles off
my intended route. It was mileage that would wear on me all

the same but count for nothing. Hoping to avoid being stuck in the desert with nothing but energy gels again, food was first on my list. My cart pushing skills have dramatically improved. The dedicated shopping Moms have nothing on me I thought while taking a tight right hand turn and reaching for a jar of creamy peanut butter.

The desert sun rose on day sixteen, and I headed northeast on Route 66. The road arched over the more direct Interstate 40. My legs were beginning to feel stronger with every mile covered. The day went by quickly as I knocked off five-mile chunks at a time.

Late afternoon found me walking down Main St. in Hackberry. It was a ghost town, like so many others along Route 66. They had all been by passed, forgotten, abandoned with the addition of the more efficient interstate. It was sad, almost like the road and the people along it had gone into a deep depression. A playground swing at a boarded up one room schoolhouse swayed back and forth in the stiff canyon breeze. I altered my path to go around some tools that were scattered in the road surrounding an old Ford pick up truck. The wind pinned trash up against a chain-linked fence that ran around a run down trailer house. The only sign of life was a young girl who kicked a soccer ball one last time before running inside. There was little reason to think I would find housing in this town.

About a mile out of town, barbed wire fences kept the invisible cattle away from the gravel road and towards the hills. A couple of small trees stood at the foot of a steep incline a few hundred yards off the road. It ended up being a good place to string my tent hammock and get out of the wind. As I started unpacking Ruby, a truck drove up. A large man jumped out and slammed the door behind him as he headed straight at me. I have to admit that my usual instincts in these types of situations are to just run in the opposite direction. However, ignoring my initial reaction and taking a quite opposite approach I jogged up to the man with an out stretched hand and a smile. "My name's Abe Clark, I'm running across America." His straight face remained motionless as my hand slowly sunk back down to my side. He had a pistol tucked in his dirt covered jeans and I realized he was not visiting to sing songs around my campfire and make some s'mores.

"You know whose land you're on?" were the first words out of his mouth.

"No I don't. I have no idea." Trying to explain myself I continued, "All I have is this tent hammock and was looking for a few trees to string it up for the night. I'll probably be gone before sun up." I felt the need to fill the silence he left. "I don't mind moving it now though, I can just keep going."

He finally spoke again, "My daughter saw you come through town, and I followed your tracks out here. This is my land from the top of that ridge clear down to the railroad tracks." As he turned to head back to his truck he said "I guess you can stay the night, just make sure your fires out"

I awkwardly yelled, "Thanks" and waved as he drove off over the rough desert floor. When he was gone I rolled my eyes and said sarcastically to myself, "Wow, thank you so much, sir for allowing me to stay the night in your coyote filled canyon for FREE! I guess this is my welcome to the Wild West."

Day 16 – 29 Miles

"I have been sitting out here in front of the warm fire for hours gazing up at a crystal clear sky and blanket of endless stars. The only trace of humans is the distant glow of a few Hackberry house lights. There is no cell phone reception here in the canyon and a few minutes ago it dawned on me that this is a rare moment. Everyone has his or her opinions about God and religion. This moment almost seems as if I am meeting God in heaven, face to face for the first time. There would be no one with me, none of my worldly possessions. I would not have brought a list of all the good things I have done or the bad. It would just be me kneeling in front of God in awe. Under a canopy of stars, completely alone in a silent Arizona desert, I have no questions."

The coyotes gave me a 4:45 a.m. wakeup call, and I decided to build a fire and wait out the last couple hours of remaining darkness in warmth. I had gathered enough wood last night for such an occasion. After spending so many nights out in the cold desert, I picked up on a few things that made life bearable. Wrapping the emergency blanket around my shoulders I stood over the fire. It acted like an oven with me in the middle. The last few cold hours of darkness crawled by and I left at the first signs of daybreak.

Although cold at night, the sun would bring the temperatures up to 70 degrees during the day. Unfortunately the warm sun on my face was not the only external warm feeling I would have as I ran down the road that day. My stomach shifted, causing my eyes to widen. Immediately squeezing my butt cheeks together I attempted to keep unknown forces from escaping. My head spun like a swivel as I frantically searched for a place to dart off the now busy road and rescue my only pair of running shorts. The shorts barely had any fabric to begin with, but now were in grave danger. There must have been some kind of road construction crew that released an endless line of cars to drive by and gaze at the diarrhea hit, homeless looking man, who unfortunately was me. The road was cut into the side of a large rock formation. A guardrail on one side of the road attempted to stop cars from plummeting to the rocks far below, and a rock wall on the other side shot straight up like a sky scrapper. There was nowhere to go. As I backed into the rock wall, dragging Ruby with one hand, I suddenly felt a strange sensation. It felt like someone had just poured a warm pot of chili down my shorts. The harder I squeezed, the more violent the eruption. My adrenalin was now rushing as I realized the severity of the situation. Scrambling to reach a place of privacy, I tripped over a sage bush. My face planted firmly into the ground, the sudden jolt caused my ballooning shorts to spill the warm chili up my lower back. I slowly closed my eyes and let out a big sigh. The world seemed to come to a sudden stop, as I lay there motionless.

In general, I believe it is important to always look on the positive side of things and believe in yourself- to be able to overcome. However, every once in awhile a situation comes along that you and everyone around you can only describe with that one simple pathetic word. Fail.

It is almost a relief when you finally reach that point. Life is just so much easier when your pride, dignity, and self-respect have been stripped away. In a sense you have nothing to lose. Holding on to that enlightened thought, I simply flicked off my running shoes and socks pulled off my shorts and began to wipe off my butt and back with the unaffected part of the shorts. A semi truck blasted his horn and I looked up from my task to see a bearded man laughing as he drove past. It dawned on me that standing butt naked on the side of a highway with cars zooming past is probably not a good idea. I stood Ruby up and attempted go about my business more discreetly. I had recently lost my lost running pants to the road so was forced to wear the pure white, ten dollar Wal-Mart spandex pants I had picked up for some extra warmth back in Kingman. After cleaning my backend off the best I could, I emptied out the last few slices of bread and dropped my soaked shorts into the only plastic bag available to me. They were my only pair so I would have to try salvaging them at a gas station down the road.

Day 17 – 23 Miles
"Jogging down the road today I was contemplating if people stared longer at me wearing my blinding

white spandexes or when I was in the ditch wearing nothing at all. Either way my stomach felt awful, my body lethargic, and all pride vanished. It was a long day. Finding strange things alongside the road seems to be a daily occurrence. Today Ruby rolled over an FBI agent's identification badge. He was the chief of police for the Hualapai Indian reservation. I think I might be losing it a bit. After I picked up the two fold I kept flipping it open while yelling, "ON THE GROUND THIS IS THE FBI!"

A few miles down the road a police car passed and I flagged him down to hand over my treasure. Taking his time he got out of the marked car and approached me, "What's the problem?" The young officer asked.

I answered "I found this police badge on the side of the road a couple miles back, and was hoping you would know who to give it to."

Flipping open the badge and pulling out the ID he took a look at it and laughed, "The chief has been looking all over for this, he lost it two weeks ago."

"Well I'm glad I was able to help." I said as he continued to study the badge.

A moment later as he put the badge in his pocket a puzzled look came over his face as if he had just first noticed my mode of transportation. "What are you up to out here anyway?" Only being a little over two weeks into my run I still hadn't perfected a short response to the simple question that made any sense.

I said the first thing that came to mind "Oh I'm just running." His stare caused my mind to go blank. Under his peach fuzz attempt at a mustache, his lips tightened and his gaze penetrated my soul. All of the sudden an overwhelming feeling of guilt came over me when I realized that this young officer probably had a one track mind on running. I could almost watch his thoughts form... "People never run to something they are always running from something."

Tucking his chin into his throat he broke the awkward silence with the lowest voice he could muster, "Do you have an ID on you?"

Snapping back to reality I tried to match his serious mood, "Yes, yes I do." Turning to Ruby I knelt down behind her to dig my Wisconsin driver's license out of her back pocket.

Causing my heart to skip a beat he quickly blurted out, "Stop! Step away from the baby jogger," a phrase I never thought I would hear. His hand hovered over his gun as if he was ready to draw on me. Tucking his chin back down into his collar he fired off several demanding questions, "Do you have any weapons in the baby jogger?", another question I thought I would never hear asked. "What's in the black bag? Is it mail?"

Thoroughly confused I answered all the questions the best I could. The answers satisfied him and he let me continue the process of retrieving my ID. Snatching it out of my hand he returned to his police car to run the license on his computer. I stood next to Ruby on the gravel shoulder with my hands hanging limply at my side. Looking down at my bright white long sleeve t-shirt and matching white spandexes I wondered to myself it they were see through. Not so much the shirt but the skintight spandexes. My thoughts were interrupted as he stepped out of his car to return my license. Hoping he would not search Ruby and find my ten-inch bear knife and sandwich diarrhea bag I met him half way. He apologized saying, "I thought you might have been someone we were looking for." With that he hopped back into his car, made a u-turn and drove out of sight.

Ten miles later another police car turned on his lights and pulled over in front of me. This one was an unmarked black SUV. Before he could introduce himself I recognized his face. It was the chief of police, the man whose ID I had found. He turned out to be a lot less jumpy than his young understudy calmly explaining, "I was hoping to find you out here and ask where you found my badge. I lost my wallet along with it and was going to go look for it."

Directing him the best I could I replied, "It was right on the shoulder of the highway across from a school about ten miles back." He was grateful and before he left told me "Someone has been breaking into over fifty cluster mailboxes around the area, mostly over by Bullhead City, anyway keep an eye open. There is a $10,000 reward. The policeman you flagged down thought you looked suspicious with your bags and everything." With that he left to go search for his lost wallet, while I continued the opposite direction into Peach Springs. Ruby and I kept a weathered eye for any suspicious looking mail carriers.

Ahead was my longest day yet, 37 miles, from Peach Springs to Seligman on route 66. After a good nights sleep and continental breakfast at the local inn, I was ready to go. The first 15 miles flew by in about two hours and 35 minutes despite some large hills.

Day 18 – 37 miles
"Mile markers lined the car free highway today so I took the opportunity to try keeping track of my mile splits. I started the day in Peach Springs at mile marker 103, at mile marker 118 I started taking splits on my watch. The following are the splits to my last 22 miles of the day. 11:14, 8:12, 9:54, 10:05, 15:40, 15:38, 13:25, 10:56, 14:41, 8:38, 7:50, 8:14, 14:34, 8:18, 8:28, 14:34, 12:04, 7:42, 7:43, 14:23, 9:25, 8:47 The splits are accurate but include a few bathroom and photo stops. On most days my running method is as follows... I begin by running as long as I can- today it was 15 miles. After that I go into a dead man's run where the constant changing of gears maxes out all of my muscle groups. For example, walk a mile, run two or three miles, walk a mile then run two or three miles throw in a mile surge. Because I am pushing a 70lb. baby jogger, my splits will

usually reflect the terrain I'm crossing. I covered the 37 miles in 6 hours and 35 minutes, which averages 10:30 a mile. Eventually I hope to eliminate the walk mile and substitute in a slow jog. Although, speed is not a concern, I am just looking to make it from point A to point B, day in and day out."

The wind was whipping today. Black clouds spotted the sky, dumping every form of precipitation possible. I spent most of the day trying to outrun rain clouds and hiding in culverts when a passing cloud unleashed its painful hail pellets. My day's efforts were rewarded when very nice couple welcomed me into their home in the town of Seligman. She was the principal of the small Arizona high school and showed her school off by inviting the assistant and head coach of the girl's basketball team over to meet me. Over dinner the four of them simultaneously told me all about their amazing run in the Arizona state basketball tournament. Finishing 4th place was a feat that the small town had rarely even dreamt possible. There weren't any seniors on the team, so returning for the title the following year was apparently on the whole towns' mind.

The following morning the clouds dissipated leaving a bright sunny day. There is a lot of open range out here and if you're not on a main road it's common to see some sort of livestock wandering around. As I came over a small hill, a hefty black bull cow stood right in the middle of the road staring at me. I didn't see the animal at first and was surprised by his presence in the barren landscape. His horns protruded out of his thick scull like knives. My slow jog had been frozen in place as we stood staring at each other. I looked him in the eyes, but noticed his gaze fell slightly to my right. It dawned on me his trance was on the bright red three-wheeled stroller called Ruby that rolled along at my side. Raising a hoof he slammed it back to the ground, his nostrils flared letting out a shot of air. Apparently this was his road, and Ruby and I were not welcome. Slowly backing down the hill, we decided to make a short off road loop around the steaming bull and just avoid the situation all together.

Day 19 – 22 miles

"With a mile to go until Ash Fork, the road just ended and left me in a cattle field that had a barbed wire fence surrounding it. I have quickly developed a pet peeve for barbed wire fences. When I did get Ruby and myself over the fence, what I thought was a road on the other side turned out to be railroad tracks and another barbed wire fence. I followed the tracks down to the interstate bridge and repeated the process of unloading Ruby and throwing her over. Finally back on track, I made my way to the Ash Fork Inn. For only $29 bucks, they gave me four walls, a roof, a two channel TV and some dirty towels. I continue my hotel ritual of emptying the ice machine into my bathtub and numbing myself from the neck down. I lay shaking in the cold tub for nearly an hour just watching the steady drip of the faucet. I feel alone and trapped in an adventure that only has one way out. The door is over 2,000 miles away. The initial excitement of the run has worn off and the miles feel endless and the hotels or camping is lonely. I don't know how much longer I can hold out like this."

The only road to Flagstaff and my stop for the night, Williams, was on I-40, so for 18 miles it was Ruby and I vs. the semis once again. In the distance snow capped mountains began to peak over the pine trees. Having not seen trees in quite some time the change of scenery was welcoming and beautiful! From Ash Fork to Williams the elevation rose about 3,000 ft. It was either flat or uphill the entire day. Patches of snow lay hidden in the shade of tall pines while the air got noticeably cooler. My legs burned as the uphill miles started to add up.

At the top of a large rise I turned around to take in the breath taking view. Pushing Ruby to the side of the road I grabbed my camera and climbed a small embankment to take a picture of the valley. The wind had been gusting down the mountain all day adding an extra punch to the difficult route. Suddenly my heart skipped a beat when out of the corner of my eye I spotted Ruby rolling backwards back down the hill in the middle of I-40. Apparently I had forgotten to lock the foot brake, and the wind had simply set her into motion. My adrenaline surged as I quickly glanced over my shoulder to see a pair of semi trucks barreling down the mountain at 75 mph. Ruby was at least 40 yards off and quickly gaining speed. Darting into the middle of the freeway I reached my top speed within seconds. If any one was watching the scene unfold from the opposite lane this is what they would have seen. A baby jogger rolling backwards down a mountain in the middle of a freeway with a man dressed in white also in the middle of the freeway sprinting at top speed to catch it with two semi trucks quickly approaching.

Life went into slow motion for a moment as I glanced over my shoulder again at the approaching semi trucks. Then everything sped back up as the front semi blasted his horn. I reached for Ruby, grabbed her black handle bar, and dove into the grassy median. The gust of wind caused my eyes to squint as I watched the trucks fly past. My heart was racing thinking of how things could have ended.

Ruby's close encounter with death heightened my attention to the road kill that lay along the road. After the incident I saw four dead elk in the ditch that had attempted the same reckless behavior as Ruby.

Williams was a quaint little mountain town thirty-five miles west of Flagstaff. The elevation was 6,700 and snow capped mountains towered over it. The town was the home of the Grand Canyon Railroad, which used to be the only way for people to get out and see the Grand Canyon. It has since become a tourist attraction that transforms into the Polar Express for Christmas. Children from all over the area come in their pajamas to ride the train 15 miles out of town to Santa's Village.

I stayed with a family by the name of Lee for two nights, and on Sunday they took me to a small Methodist church a few blocks from their home. I counted 35 people in the congregation and felt incredibly welcomed. The 100 year-old church looked pristine for its age. Currently the congregation was raising money for a new bell that would echo throughout their mountain town.

Instead of taking a day off and making one long thirty-five mile push to Flagstaff through the snow, I decided to spend the afternoon running to Parks and split the mileage up. The Lees were a kind family and offered to pick me up and let me stay a second night. Ruby stayed behind to recuperate from her brush with death so I continued on with only a water belt, the GPS and my cell phone.

Snow began falling in large wet flakes and made the small mountain town into the perfect winter scene. It was strange to go from the desert to a snowstorm in a matter of days. The snow reminded me all too well of my Wisconsin training runs leading up to my transcontinental attempt. I overshot my first turn by a half mile. When I got back to the turn I realized why. The road was basically non-existent. If there was a road there it was under three feet of snow and behind a barbed wire fence. This road connected up with a dirt road in a mile and a half, and I decided it would be safer to trudge through the snow than be on the slippery I-40.

I jumped the fence and charged into the snow. The snow was fairly firm, but every couple of steps, my leg would break through the surface and I would fall in over my knees. It was slow going at best and extremely exhausting. I lay down in the snow and listened to my heart pumping, my body not wanting to go on. Finally I forced myself up and about a mile later I came to a creek that was flooded with run off from the mountains. Looking for a thin place to make the jump I approached the snow ledge. As I stepped out, it suddenly gave way; and I splashed into the icy mountain stream up to my waist. The shock of the half frozen stream knocked my breath away. Instantly soaked, I scrambled for something to hold onto. Pulling myself to safety I thought, "At least now I had no reason to not just walk across." Once on the other side I picked up the pace knowing I had to keep up my body temperature up if I was ever going to make it to my check point.

Finishing off the non-existent road with one more hop of a fence, I landed on a freshly plowed, muddy road. The road lead back onto old Route 66, and I could see my destination a quarter of a mile away. The wind made my wet body turn to ice. I turned off my brain and forced myself to keep a pathetic jog going. The black top of Hwy 66 turned to gray then to white as the snow accumulated. I could no longer feel my body as I burst through the doors of the Park's Gas Station. Warmth.

Camping along Route 66 in Arizona, south of the Grand Canyon. At this point I had already gotten rid of all my water bottles and hydration bladders and simply drank water out of a gallon jug, filling it whenever I found a faucet. The nights in the desert got much colder than I expected and I usually attempted to keep a fire going throughout the night.

CHAPTER 4
NAVAJO

The snowstorm had passed and sun made the freshly fallen snow sparkle like diamonds. The roads were plowed and for the most part, I was able to put together a good route to Flagstaff. The peaks of the mountains disappeared into the clouds leaving me to guess their sheer magnitude.

Day 22 (Flagstaff) 25 miles
"The Stumps lived on the east side of Flagstaff right off Hwy 89 which I will later take north to Tuba City. I made decent time today and my legs felt strong. There is something about a mountain town that makes the homes look fairy tale like all nestled in-between the white peaks. Approaching the Stump's house I heard a sharp whistle and looked up to see a man waving. They were expecting me. After introductions Joyce put on an

early dinner for me and we feasted on roast beef, rolls, vegetables, rice and more! While we ate Lowell told me about the area. He said that the Flagstaff Mountains used to be volcanic so the area's wealth and success is due to the large amounts of cinder. They have red and black cinder for making cinder blocks. Supposedly the black cinder is more expensive due to its strength. After dinner they prepared for their weekly visit to the nursing home. Joyce played a variety of hymns and praise songs on the piano for the residents to sing along to. The Stump's had a guitar so I went along and did my best to keep up with Joyce's bouncing piano fingers. We had a great time."

The days ahead started to mesh together. The randomness of the journey became my daily norm. The unexpected was becoming expected as I dropped out of the snowy elevation of Flagstaff and into the Painted Desert. The Stumps had asked me if the elevation and thin air had affected my running at all. To be honest, the thought didn't cross my mind until they mentioned it. The next day I did notice it. It was as if eating wafers for dinner; no matter how many you eat, they don't leave you completely satisfied.

Thankfully I did not have to live off wafers; instead, Joyce had given me the rest of her Peanut Butter Nut Cookies, and at mile six I decided to award myself one cookie every two miles, saving the only White Chocolate Chip and Almond Cookie for mile 26. The cookie had rave reviews from Lowell, and was the only one of its kind he was willing to spare. Along the way I was able to push Ruby to a new timed mile record of 6:30. I stopped the clock at 3:55 after mile 26 and sat on the side of the road eating my reward, waiting for my grandparents.

They had decided to leave the comforts of their Arizona winter home in Sun City to drive up and spoil me. It was hard to complain. They seemed to have learned a few tricks in life as we sat in Buffalo Wild Wings laughing and complaining about the dry pork. I couldn't believe it as they both walked out the door each with a free sandwich. Most of all it was just great to see a familiar face.

Running straight north into the Navajo Indian Reservation, the scenery began to change once again. Bright, jagged red rocks stretched out in front of me, bewildering my imagination. The dark blacktop and vibrant blue sky over saturated my panoramic view with color. This nation is made up of thousands of landscapes; my horizon is forever changing; my road is never ending. A tree stuck up out of the painted dessert floor. I pondered what its survival story was. How its seed was able to take root and survive the unimaginable? What makes it so different than the thousands of seeds that have tried and failed or simply looked around at their circumstances and just gaven up? It is easy for a tree to grow with fertile soil and a balanced mixture of sun and rain. However, it usually goes unnoticed in a forest of similar species. There is something strong and silent about a lone tree standing in the mist of a barren desert scene. I stood in its shadow for a moment and watched a car whiz past, the eyes of a child in the backseat locked with mine. She got up on her knees and turned in her seat to peer out the back window. Our present horizons eventually tore us apart like paper. I often wondered if she thought she was staring at two trees.

Thirty-three miles later I came to Tuba City one of the largest settlements on the 26,000 square mile reservation. The Navajo Indians have occupied the reservation since it was established in 1868. Every five to ten miles I would see a booth set up on the side of the road with a lady trying to sell her authentic Navajo Indian crafts to the tourists that came to see the famous painted desert.

Day 27 – 38 miles
"My diet seems to change with the terrain and culture I

run through. The Indian tacos I had last night did not mix well with my daily energy bars and gels. Less than a mile into my run today I had another messy situation on hand. Outside of Red Lake there is a national monument called Elephant Feet. Unfortunately behind the left foot of the larger than life elephant feet was the only immediate and somewhat private location to take care of my runs. I hope none of the tourists look behind the left elephant foot today. By my fifth stop, I got pretty good at anticipating the sudden eruptions and do take a small amount of joy and satisfaction in the fact that I am no longer completely ruining my running gear. Despite my muscles and insides completely falling apart my eyes were fascinated on the extraordinary new landscape. Patches of pure white snow lay on the cliffs contrasting the vibrant red rocks."

I reached Tsegi Canyon an hour before sunset and stood and watched the sun's rays find holes in the cumulonimbus clouds as they shone down on the magical painted canyon. An old sheep dog slowly walked up and sat down next to me. I enjoyed the company and found comfort in his innocent approach. You could tell in his old age he learned to stop and appreciate when God got out his paintbrush and went to work. When you're the only one there watching a sunset, it becomes a very personal moment with God.

The next day a small red car with rusted fenders slowed down to my snail pace of six miles per hour. While I was still running he shouted out the window, "You must be Abe Clark?"

I nodded and with a smile replied, "I am." He was a happy man overflowing with life and apparently on a mission to help me in every way that he could. Three similar visits later he had reserved a hotel room for me in Kayenta and arranged to pick me up for their Sunday night service. His excitement brought my moral up a few notches and the day's miles drifted by unnoticed.

As I trotted into Kayenta, I could see the unique rock formations of Monument Valley to the North. I was retracing the footsteps of an American legend, Forrest Gump. Monument Valley is the famous last shot where Forrest, played by Tom Hanks, stops running and says, "I'm pretty tired now I think I'll go home."

After I got settled in, Pastor Billy came to pick me up for church. The church only had four members when he started two years ago and today the congregation has increased to around 85. About 12 of the members were there for the night service. They sang half of their hymns in Navajo, and I began to wonder if mouthing, "watermelon" worked the same way with the Navajo language as it did with English. There were two long florescent lights above the pulpit and one flickered the entire service until it finally went out in the middle of a lady's special song. Pastor Billy called me up next to introduce myself and talk about the run and Living Water International.

My opening line was, "You know some musicians pay a lot of money for lighting effects like that." The 12 elderly Navajo citizens silently stared at me without giving me the slightest bit of grace in my attempt at humor. Deciding it was wise to cast myself as the bunt of my lame jokes for the remainder of my talk panned out well. The seriousness of my mission for fresh drinking water was a problem that they could all relate with. By the end of my talk, they were visibly intrigued with my quest. Afterwards, we sat around the wood stove they used to heat the building, ate fry bread and traded stories. Pastor Billy told me that the elders are very hesitant about any white people that pass through the area. The Navajo people are kind, proud and generous after you break through the shell of caution. I was grateful for their hospitality.

I crossed the desert valley that surrounded Kayenta and continued on a northeastern route. Snow capped mountains began to reveal themselves behind the red rock formations as

I inched towards the Rockies. A Navajo woman led her sheep across the patchy vegetation dirt landscape. Scenes like these are helping me grasp how big and diverse America is. There is a good chance the white collar worker from California will never meet the Navajo Shepherd only one state over. I guess that's the beauty of America. A man zoomed by with a jacked up truck and a trailer full of four wheelers and yelled, "Ride a horse!"

I gave him creative points at first, but of course it was the only subject I thought about for the next five miles. I decided Ruby was better than any horse. She rarely misbehaves, never asks for food or water and is always right in front of me no matter how fast or slow I'm traveling.

Anticipating the joy of reaching a town usually ends in disappointment. The towns out here are very small and run down. I managed to find a place to sleep indoors for the night. It was an old boarding house now used as a storage facility for the nearby school. It was full of random school items: desks, computers, books. I was lucky to find an old mattress that seemed to be the victim of an angry child from the 1980's. I dragged it into the shower room, which appeared to be the cleanest room. The area gets a lot of dust storms; dirt seems to have accumulated under the windowsills in small mounds. As I wrote in my journal, I started to sweat profusely. Wearing nothing but my running shorts I was still sweating more than a midday run. A drop of sweat trickled off the tip of my nose and splattered onto the inked page. I had it. Standing up I searched for ventilation but soon realized my reading light was actually a heat lamp for the showers. With the flick of a switch the temperature dropped, and I fell fast asleep in the most unlikely of places.

Day 30 (Red Mesa) 31 Mile Run

"There have been an increasing amount of horses along the road. I don't know if they are wild or not but they appear to be. They often run in groups and it's quite a sight to see them thundering along over the wide-open ranges with white capped mountains for a backdrop. From watching the *Black Stallion* movies, I always thought it was a bit ridiculous that a group of horses would have a leader but sure enough if you sit there and watch them you can usually pick out the head horse. I still can't believe how far you can see out here in the West. This place is huge! It's a far cry from the tight knit north woods of Wisconsin."

The long road through the desert only urged my growing anticipation, of what the next town would hold. Some locals told me the only modern building in Red Mesa was a Four Corners area clinic serving the Navajo People. It sprung up due to some government influence and funding. Even though the town was still six miles off I could see it rising above the dull brown sand. The orange and reds of another desert sunset lit my remaining miles. Houses made from a government cookie cutter surrounded the surprisingly large new hospital.

Walking through the empty parking lot I feared it would be too late to find anyone to talk to. Suddenly a straggling nurse exited the building and began to cross the lot. Putting my hand up to wave, I called out, "Excuse me, Hi." I had learned to be very bold in explaining myself in situations like this. "My name is Abe Clark, I am running across America, and had called ahead about possibly staying in the EMT holding room for the night. Do you know anyone who I could talk to about that?"

My statement took her back a moment and she snapped into action muttering something to avoiding the awkward silence. She put down her bags, dug out her phone and declared as if the idea had just hit her, "I think I might have some phone numbers that could help you. Would some phone numbers help?" She stared at me with wide eyes, rapidly nodding her head as she waited for my reply.

"My phone doesn't get service out here do you mind calling on yours?" She agreed and tried several numbers. Unable to find any answers for me, she apologized and continued on her way home. A lit overhang jetted out over the front doors of the hospital. Being the only light place around, it looked like a good place to rest and wait for events to unfold. A large cement bench ran next to the doors so I got out my sleep pad to rest. Lying stretched out on my back I crossed my arms and stared at bugs circling around the bright light. I fought to stay awake but my eyelids gave into exhaustion.

A voice from the parking lot called out. I shot up and looked at my watch. An hour had passed and as my eyes woke a security guard came into focus. Unable to comprehend his words I said back, "What was that?"

"I'm guessing you're the guy looking for a place to sleep for the night. I got a message saying the paper work didn't go through for you to stay in the EMT room, but there is a young couple that lives across the lot here who would be happy to house you."

With an outstretched hand he said, "My name is Don. I'm in charge of security around here. I can take you over there if you like."

"That would be great," I said shaking his hand.

As I picked up my things Don rambled on, "You will be staying with Tom, he has a few adventures of his own you should ask him about."

Sure enough Tom had some interesting tales to tell. Upon graduating from college he made a pact with some good friends to meet for a South American bike tour a year from the day. Tom said, "I was pretty surprised when the year was up and every one was actually ready take our mountain bikes down to South America. We spent a full year riding around with no particular destination in mind."

We talked for hours, trading random stories of adventure.

Eventually he noticed the late hour. "I have to work in the morning, but you can stick around and sleep in if you like. Just lock the door when you leave."

When I woke, Tom and his wife had already left for the hospital. They had left a note saying how fun it was to have met me and to take whatever food I needed. I headed out hoping to cross into New Mexico but had no set plans on where I would stay. Wispy white clouds marbled the bright blue sky. I was only 20 or so miles from the Four Corners Monument, and I could clearly see the mountains of Colorado to the northeast. I stopped at the Teec Nos Pos trading post for an ice cream sandwich and sat in the shade returning zombie like stares from tourist on a bus heading to the Four Corners.

It felt both exciting and exhausting to start a whole new state. I'm not one to believe in superstitions, but it was a little strange how a black cat chased me across the New Mexico border meowing like I stole his only can of Weruva Paw Lickin' Chicken. Only time would tell if Ruby and I would survive the early March crossing of the Rocky Mountains. There was no way of getting around the Rockies, and in the back of my mind, I knew that the challenges I had faced so far would seem minute in comparison to the road ahead. Climbing the elevation alone would wear me down to the bone but the endless upward climb was not what I feared the most.

Another 16 miles on top of the 25 I had already come that day would bring me to Shiprock New Mexico named after a giant rock formation that resembled a ship outside the town. With the sun getting low, God began to paint another impressive sunset. Wispy clouds turned into colorful brush strokes that eventually left me smiling in the dark. The moon hung like a hammock in the sky. With four miles left, the darkness turned my peaceful run into blinding headlights and uncertain footing. The air was cold now and my short sleeve shirt and shorts were no longer ideal running gear. From behind something charged straight at me, the sudden approach caused me to spin around and brace myself. It was a wild horse, apparently attracted to the blinking lights that littered Ruby's frame. We stood motionless in the cool night staring each other down. Our souls understood each other's- we were wild, and we were free. Just as suddenly as we stopped we both took off again faster than ever. His heavy hooves beat against the helpless ground. Powerful breaths from his nostrils made the animal's efforts visible in the moonlight. We flew through the desert night, our long hair floating on the wind. We embraced the darkness, shut out the world, completely locking us into the moment. My senses heightened, and I felt completely in touch with my body as it sliced through the night. In that moment we weren't running to get anywhere, or for any reason. We were running just to run, and to us, that made perfect sense.

Day 32 (Farmington) 27 miles
"Pastor Bill of the Assembly of God church in Shiprock was a big help last night. He took me out to McDonalds to get something to eat and meet a few people. The director of foreign missions was there to talk about donating money to Living Water. After a 41 mile run it took a double cheeseburger, large milk shake, chicken wrap and two large Powerades to satisfy my craving stomach. I stayed in an old trailer that was next to the church. They used it for Sunday school and youth group meetings. There was no sleeping furniture so I chose a small room at the end that looked happy. Children's artwork hung on the walls explaining why they liked attending church. I put in a *Veggie Tales* tape and fell asleep on the colorful alphabet rug listening to Silly Songs with Larry."

A closed seasonal cafe was the only building that marked Navajo City was as a town at all. The rundown place was in desperate need of a paint job. A few trailers were scattered about in the small pines, and I knew there would be little hope of human interaction. Reevaluating my plan for the night, I decided to move on towards the wild national forest instead of stay in the non-existing town.

Barbed wire fences lined the side of the road making it hard to find a place to get off and set up camp. Eventually I came across an actual house and was surprised to see a person outside working in the yard; I knew it was my only hope. Drumming up my courage I headed down the quarter mile dirt driveway. As I approached, his pack of horses ran up onto the driveway and blocked my path. My steady forward motion sent them scampering off into the yard as I grew near. Watching the horses gallop around the water hole, a dog jumped up from his lazy resting place to come check out the commotion. Unlike the horses, the dog was not afraid to thoroughly check me out. Between barks he continually jumped up on me sniffing my clothes like a vacuum cleaner. I had a hard time deciding if he was growling at me or that's just how his face looked. Some of the dogs out here on the Navajo Reservation are the scraggliest looking things. The man came over and I introduced myself. I kindly asked if I could camp in his yard or the barn for the night. Of course I wanted to be in the house but that was something he would have to offer. He thought for a minute and said, "I guess it wouldn't hurt anything, maybe you could set up somewhere over there by the shed. Do you have a tent?"

Taking a deep breath I exhaled, "Yeah it's actually a tent hammock so those post under the overhang would probably work pretty well." His collection of used vehicles and tools led me to believe he had been there a long time.

"Well, I have a few chores to finish up, I'll stop by before I go in."

I had cut my set-up time down considerably since my first night out in the open, and within a half hour, I was sitting in the hammock writing in my journal. The man drove up on his four-wheeler, and I did my best to listen to his long rant about everything that was wrong with the government.

Coyotes howled and yelped at the dark landscape, and my scraggly new friend did his best to discourage them by adding his own bark to the choir. I thought to myself, "How can I blame him, I've seen people attempt the same thing." Even though the day had left me exhausted I still fell in and out of sleep, either being woken by the howls or the cold. I felt paralyzed and helpless in a frozen cocoon unable to save myself from the night's misery.

A bright sun blinded my eyes; unable to see where I was going, I fell to my knees. The sun beat down on my flesh sucking the last few drops of liquid from my body. Like a sightless man, I scrambled about in the sand searching for my jug of crystal clear cold water that I kept tied to Ruby. As I fell back, I grabbed the jug of water laying face up in the middle of a dried out cracked desert as I placed it upside down over my face. Nothing came out. I shook it once, then again and again, yet the water hovered over me defying gravity and mocking my thirst. With the last of my energy I threw the jug as far as I could; it came to a skidding stop and began to bubble out and disappear into the dust I was becoming.

I jolted in the hammock causing it to sway back and forth. Taking my hand out of the sleeping bag I reached over and felt my side. It was soaking wet. I had peed right through every single layer. Too tired to move I closed my eyes again. The next time I woke my body was shaking uncontrollably. The wet pee had caused my body temperature to drop leaving me in a hypothermic state. I ran in my sleeping bag to warm up and regain control of my body but my efforts only lasted so long. Checking my watch for the tenth time in the past half hour I decided dawn was in reach and I crawled out of the hammock to pack up Ruby for another day on the road.

Day 37 (Dulce) 47 miles

"I was so cold last night I had to run inside my sleeping bag just to stay warm. It was a rough night and I headed out as soon as the sun's glow began to peek over the mountains. I wish I could harness the joy I felt to see the sun come up. It would be great to feel such pure excitement over a simple sunrise every morning. I don't know what is ahead or where I will stop, I guess the day will reveal everything in time."

Ten miles into my run a man pulled over in his pickup truck. Rolling down his window he said, "Are you the kid who's running across the country?"

With a half smile I replied, "Yes, I am, at least I'm heading that way."

My nonchalant attitude must have concerned him. "I heard about you on the radio yesterday, you know this is primitive country out here right?"

I looked around, "Yeah, not much out here, is there."

"Don't know where you're planning to sleep to night but I would pick your site carefully, you will be on Apache land soon and if they find you camping on it they'll haul you off to jail. Either way Carson National Forest is full of Mountain Lions and there's a snow storm forecasted for tonight."

I shrugged and looked at Ruby, she remained expressionless. "I'll figure something out." There was no question about his knowledge of the country I was in; he had worked the oil rigs that were scattered about the rugged hills for years. He drove off shaking his head leaving me with thoughts I could have done without.

The afternoon wore on as the skies turned overcast, clouding my mind with the worries that the man in the pickup truck had forecasted. The one ray of hope that he did manage to let me in on was that the nearest hotel was 37 miles from where we were. Make a detour to the hotel in Dulce would mean adding 30 plus miles to my route. However, the thought of a warm place to sleep for the night weighed heavy in my mind providing the reality that running another 32 miles that day was even possible.

Twenty-five miles later I sat on a cold steel guard rail that guided the twisty mountain road. The black clouds covered the sunset, and day turned to dusk without the pleasure of color. I had used every running mind trick I knew to get me to this point and all it did was leave me seven miles short of warmth and security. I have never felt more exhausted than I did now. Snow began to softly fall into the canyon, twirling and dancing on the cold pavement. I reached into a pocket of Ruby that I rarely opened and pulled out the picture that Bill Ash had given me back on the warm California beach where I had started. Holding the image that was still frozen in time, I noticed the boy's dark skin and how it contrasted my white hand and the snowflakes that fell on the photo. We were polar opposites in more ways than just color, yet we were still connected in separate moments that formed a single story.

I stood to my feet, every joint from my waist down felt compressed and unable to support the weight above. Placing the picture on Ruby's canopy I continued to stare at it, slowly starting to move forward, placing one foot in front of the other. I knew it was up to me to finish my part of our story. My slow steps broke into a painful jog, and my face winced between exhales. I stopped thinking about the aches and pains and the seven miles ahead and focused only on the picture. It was ironic how the run was supposed to be about me saving him, yet as darkness began to swallow the mountains, it was him who was saving me.

Above - Running into Twentynine Palms, CA. My older brother set up some housing for me with a few Marines that worked on the base in town.

Below - Camping in an abandoned cattle pen, the space blanket wrapped around the hammock for warmth.

CHAPTER 5
BLIZZARD IN THE ROCKIES

I awoke with my running clothes still on, which as of late was not unusual. Socks lay on the floor with the soles of my shoes attached as if someone had glued them together with blood, blister puss, and sweat. Last night I had finished off the longest run of my life. However, the unexpected additional miles did not put me any farther ahead on my route but did end up giving me a warm place to sleep for the night. Before leaving the comforts of the room, I took a few minutes to stretch. It was a habit I never really got into even running at college. My legs felt stiff and frozen in their running form. With knees slightly bent, hamstrings only allowing a three-foot stride, I began another day of endless miles.

Colorado was only a dozen miles to the north. The scenery and the small town feel were slowly changing to mountain culture. I made it to Chama at dusk and inconveniently forgot the name of the hotel that my contact, Jerry, had reserved for me. Hotel receptionists must see many weary travelers throughout their careers. As Linda sat behind her desk gnawing on a pile of take out chicken wings, I shot to the top of her list. Her chewing stopped, the half ground chicken settled in her mouth like a load of dry laundry coming to rest.

Slowly walking up to the counter I placed both elbows on it letting them take the weight off my over used legs. I looked into her eyes waiting for her to mutter a welcome. Forgetting that her mouth was still full of half chewed chicken she tried to say what I was waiting to hear. "Welco..." remembering the chicken, she attempted to swallow. I could see her throat expand, attempting to suck down the glob of chicken like a vacuum cleaner. The lump stopped halfway and a small innocent cough came out. She paused and focused hoping that would be the end of it. Suddenly without warning she darted though the swinging doors and into the back room. A chorus of hacking coughs followed, and then silence. Still leaning on the desk to tired too even flinch I pondered if I should jump the counter and see if she was okay. The seconds ticked by. Then with a grand smiling entrance she calmly opened the door and softly said, "Welcome to River Bend Lodge, would you like a room for the night?" When all was said and done I must say I came out on top, she ended up giving me the rest of her chicken dinner and directions to the cheapest hotel in town, which also proved to be my correct hotel for the night.

Jerry came over in the morning and took me to the local radio station. Chama was a cute town with the main attraction being an old train that took people over the mountains into Colorado. *Dust in the Wind* played over the radio as I walked into the station. A lady opened the studio door, confidently introduced herself and waved me in. "Hi, I'm Jenny, the DJ." I could instantly tell why she was perfect for the job; her larger than life personality and voice filled the small studio.

"Abe Clark," I replied.

"We have been looking forward to meeting you. How has the run been going so far?" She handed me a pair of ear phones to put on. I took a deep breath, "It's been going very well, I am excited…"

She interrupted, "Hold that thought I will get it on air in a moment" the song ended, she cleared her throat. "Well good morning here from KZRM Radio 96.1 Rocky Mountain Rock that was a little *Dust in the Wind* released by Kansas in 1977, again if you're just joining us we are playing songs with the word wind in them all morning long." She held out the word wind and continued. "However, we are going to take a small break from that fun and ask a few questions to a special guest we have in this morning that is literally just running by the studio. Abe Clark is hoping to become just the 15th person to ever run across America solo and without a support crew. Abe how has it been going so far? Where did you start?"

The words that came out over the speakers at a steady and unbroken flow came to a sudden halt when the conversation shifted to me. I was still trying to get used to the media attention that would sometimes pop up as I passed through a small town. Being alone for days and nights on end seemed to leave my mind in a constant state of thought that was becoming hard to turn off and make the shift toward articulating a conversation. My thoughts were alive and the stories ran wild in my mind; yet my lips were always sealed nearly forgetting how to escape at all. "Good, the run is good, I started in California… Oceanside, California."

Jenny burst back into the conversation, "Well, that's a long way from here. I've never even been to California." Her over the top personality eventually loosened me up and we had a good conversation on air. I was able to share how Living Water was bringing fresh drinking water and the love of Jesus to underdeveloped countries.

After our conversation Jenny voiced her concern for the stretch ahead. "It's over a hundred miles through the mountains to the next hotel or town with anything, and the weather's supposed to get pretty nasty. Are you sure you're going to be okay?"

Well, I have no idea, I thought to myself before reassuring her. "Yes I'll be fine."

"Okay, good. But here, take my number in case you need help anyway." I knew my chances of having cell service in the mountains was slim to none, but I took the number and thanked her regardless.

When it seemed as if I could not get any closer to the Rockies, without ripping through the picturesque backdrop completely, I saw a big church steeple about a mile off the main road. Since dusk was only a few hours away, I thought I would gamble on the old mountain chapel to provide a pew for the night before taking on the mountain pass.

After pulling on the locked doors several times, I backed away and stood in the middle of the street, staring at the white chapel. There aren't many feelings in life that are worse than having hope ripped away. However, before disappointment could completely settle an elderly lady called out from the porch of a long western looking one story building. "Hi there, are you looking for us?" A sign above the porch read Terra Wool.

My eyes dropped from the sign down to the lady below it smiling with her arms crossed. Our eyes met and she waved again. I looked over my shoulder to see if she was talking to someone else; there was no evidence of a single soul in the small tourist town besides Ruby and I. She called out, "We're the only place open this time of year... closing for the day in a few minutes if you wanted to look around quick." She waited for my smile and nod before disappearing into the building. Curiously I followed her in. It was a small wool shop full of rugs, hats, and blankets and just about anything else you could possibly make out of wool. The lady from the street was now standing behind a checkout counter still smiling. Two other elderly ladies, equally excited to see me as the first, watched me move about the shop. One mixed a fresh mug of hot tea while the other knit.

Under their watchful eyes, I felt the need to comment on something. "My mom would love this place. Is everything handmade?" I said. That appeared to be enough to set the three ladies off on a history lesson of Terra Wool and a full tour of the facilities. "Everything that we sell in the store is made right here in the shop," Olivia, the lady from the porch said. She opened two large wooden doors in the back, and they opened up to a large room. The lights were off but a big window in the back shed light over a dozen giant looms.

I was honestly very impressed with their set up and dedication to perfecting their craft. They noticed my sincerity and began asking about me. Fifteen minutes later Olivia was giving me the keys to a guest cottage they had out back. For the time being, luck appeared to be on my side.

Day 40
"A few days ago I crossed the continental divide. I was hoping the road ahead would be all downhill. I realize now that that was wishful thinking. The road ahead appears to climb into the heavens. Today I will be attempting to cross over the Taos mountains range and yet another section of the 1.5 million acres Carson National Forest totaling 49 miles to the next town. My confidence is high with the success of my 47 miler a few days ago. I hope the weather holds out."

The road climbed to 10,600 feet elevation. My hopes of covering the distance in a single day slowly began to vanish with the blue skies. As determined, as I was to get up the long mountain road, Ruby seemed equally determined to roll back down it. At every switch back turn the road would open up to a breathtaking view of the wild landscape. I climbed up onto a guardrail and gazed out over the valley in which I had come.

A feeling of complete freedom came over me. It is a rare

feeling that every adventure enthusiast craves and often spends a lifetime repeatedly searching for. This incomparable emotion feels like a soaring eagle caring your soul to the edge of heaven. When I was a young boy my brothers and I would find this fulfillment by simply walking through the 40 acre woods behind our house with BB guns. When that no longer fulfilled us we bought coon dogs and chased them for hours through the dark woods that surrounded the twisting river bottom of Oconto County. There is no clear road map to the edge of your soul, and I began to realize the more I feed this craving, the harder it was to find. The chase had to be more exhilarating than the last, the journey farther than the one before, the outcome of the adventure more unknown.

A few snowflakes drifted down to my face adding to the white frost that was building up on my whiskers. A gust of wind hit me and blew straight through my thin running jacket and bounced off the skin that clung to my visible ribs. The cold chill brought me back to the reality of my current situation. I could smell and feel the storm front coming in and knew it was only a matter of time before the scenic mountains would reveal their ugly side. I jumped down from my perch and continued into the wild mountains with my mind stuck on how I would survive the night.

It was only three in the afternoon but the dark clouds blocked out sun, making it feel as if night was rapidly approaching. As the snow picked up the swirling wind flung the flakes about on the cold asphalt road. When I reached the peak the wind seemed to clinch down on its bite carrying the fallen snow in its teeth. The stinging snow forced my eyes to close. Only an inch or so had accumulated on the road but that was all it took to slow Ruby. The slow pace and bitter cold wind dropped my body temperature, and I began to shiver. I had not seen a car pass in over two hours, and it dawned on me that I may be on my own for the night. It was still over 45 miles to the nearest town and a marathon away in the wrong direction from the

Terra Wool cottage I had stayed at the night before. Exhausted from the climb to the top of the mountain range and pushing Ruby through the snow, I leaned against the six-foot high wall of snow that lined the road. I thought of Jenny's warning back at the radio station and remembered her kind offer to help if I need it. I dug out my cell phone waiting for the frozen block to power up I surveyed my surroundings. A hundred yards off the road next to a grove of pine trees appeared to be some kind of shelter. The phone came to life with a happy jingle and I dialed the number she had given me, pressed the green call button and placed it to my ear. Silence, Beep, Beep, Beep. I looked at the phone it blinked "No service." It was useless.

Another big gust of wind hit me persuading me to check out the shed like shelter that I had spotted from the road. I threw Ruby up onto the wall of snow before scaling it, myself. My no show socks and lightweight running shoes broke through the snow with every step. Dragging Ruby on her side with one arm and clawing at the snow with the other, progress was slow. The 100 yards that felt like a mile came to a disappointing end. What appeared to be a small shed or cabin from the road was simply a small picnic table covering. The snow was piled so high that my head nearly reached the peek of the roof. Without walls the shelter was useless. I knew my tent hammock would be nothing but a death trap in this cold. It was a full out mountain blizzard by now, and there I was just standing in the middle of 1.5 million acres of wilderness at over 10,000 feet elevation in nothing but running gear.

In between thoughts of finding a desperate solution to the situation were thoughts of "How could I be this stupid?" I realized screaming at myself would do nothing and decided to snap into action and fill the two foot gap that remained between the roof and the drift with snow. Maybe I could build snow walls for the roof giving me a house for the night. With each passing moment the shivering got worse, and I realized it would probably take me a full day to close in the shelter. Even then the space would be too large for any kind of real warmth. The baseball gloves I wore for pushing Ruby were useless in this cold, and my fingers turned to icicles within minutes of digging.

I stopped working and lay back in the drift of snow breathing hard from the thin oxygen and my useless efforts. Blowing snow began to accumulate next to my motionless body reminding me of how the sand from the Pacific Ocean covered my feet back on that first day. Maybe the snow was really just frozen sand. If I closed my eyes and let it cover me, perhaps it would take me to some warm distant beach- maybe back in time to the island beaches of Door County that Kate and I spent the past summer sailing and exploring. If the cold snow would take me back there, I would close my eyes and let it.

My eyes shot open, "What was I doing?" I took a deep breath and thought. Warmth, I need warmth now. Rolling over to Ruby I reached into a pocket and pulled out a small box of wooden matches. If I could get a good fire going it would at least keep me from freezing to death in the snow and give me time to work on a shelter. The only visible firewood next to the picnic shelter was the dead branches from the pine grove. I plowed through the snow snatching up every stick I could get my hands on. The amount of wood I found was barely worth the effort it took to trudge around in the snow collecting it. However, the flame of warmth that seemed to be only a flick of a wrist away drove me like a possessed man. I threw the arm full of wood down next to Ruby and dug a small hole attempting to keep the swirling wind at bay. Breaking up some of the smaller sticks, I added a few pieces of my New Mexico map to the mix. Without hesitation I struck a match, the smallest puff of wind extinguished the flame before the blackened red and white tip could reach the edge of the torn map. I huddled in closer and tried another. This time the flickering match lasted twice as long, the small flame licking the edge of the map. Then in an instant, it vanished.

A few minutes later a dozen burnt out matches lay scattered about the nest of twigs and map. The map seemed to be slightly damp and not made of the most flammable paper so I turn to Ruby for advice. Digging around in her pockets I found a second paper solution. My journal. Half the pages were still blank so I ripped a few out and added them to my tangled bird's nest. Trying to steady my uncontrollable shaking hands, I leaned in as close as possible for the one perfect strike. This flawless motion that would bring me back to life. Time slowed, the wind faded in my ears, teeth clamped down on my bottom lip, lungs full of paused flowing air. My hand, now steady, pressed the match against the side of the box. A chorus of dry friction accompanied the well-choreographed movement. The sound of the scraping match echoed in the white noise that engulfed the white snow. A flickering flame waited patiently under the journal pages as my face hovered impatiently above. Suddenly, like a brilliant thought, the flame caught and spread growing with my intense focus. The flame spread to the branches and grew into a small fire. My hands were so close to the flame they almost held it in their palm. I carefully added sticks to my growing lifeline until my small pile of wood disappeared. I needed more wood- a lot more wood. Leaving the glow of warmth, I trudged back into the stand of pines frantically ripping off branches as I went. The fire wouldn't last long in this weather without my tender care. A few minutes later I was sprinting back to the fire with a hand full of sticks. Ten yards away from the dying fire, a gust of wind carried a fleet of fresh fallen snow that scattered my remaining burning sticks extinguishing all hope. I fell to my knees five yards from the scattered, burnt sticks, as if the gust of wind had also blown my last remaining flame of hope out. Exhaling, my eyes closed.

I felt defeated and stupid for getting myself into this situation. Having nothing but a thin running jacket to keep the wind off my wet skin, my body shook. I thought to myself, "Was this it? How long can I survive like this? I need to get out of this wind."

I opened my eyes and began to dig straight down. Like a madman after gold, I flung the snow out the hole. When the under layers of snow got too hard to scoop with my hands, I grabbed my ten-inch knife and carved out blocks tossing them out of my growing cave. I was surprised at how fast the shelter took shape. Could the snow that was threatening my life also be the answer to saving it? The farther down I dug, the more I realized that this could really work. I could close myself in completely, shut out the storm and fly under the radar until morning.

An hour later I threw everything that would help me survive the night into my snow cave. The small entry way opened up to a space long enough for me to lie down in and tall enough for me to sit up. I placed my hammock rain fly under the sleeping mat for added insulation from the snow and spread out my compact sleeping bag. Carving out one more big chunk of snow I stuffed it into the entry way and my cave went black. The howling wind softened to a distant hush and I heard my self exhale. My now visible breath floated in the still.

Exhausted from the day's run and struggle for shelter I sat back in the cave. The second I stopped moving my body resorted back to the uncontrollable shaking. My clothes were soaked from digging in the snow and felt useless on my shaking flesh. Keeping my spare clothes in a waterproof bag was probably the one thing I did right heading into this mess. I pulled out every dry article I had. A pair of jeans, two t-shirts, long sleeve running top, a couple pairs of socks and a few bandanas. After switching out all my clothes, I stuffed the wet clothes under the sleeping mat to farther separate me from the frozen ground. A glimpse of hope returned as I wrapped my over-used two-dollar emergency blanket around my shoulders and zipped myself into the sleeping bag. Consistently rubbing my shaking body was my only source of heat besides my own breath. Eventually I was able to get warm enough to stop shivering, finally giving my body a break from playing catch up. The already thin mountain air became scarce in the small cave and almost non-existent when I tucked my head into the airtight cocoon. Deep breaths slowed pausing at the bottom.

I turned in my sleeping bag searching for comfort. The slight movement caused me to lose my breath. Every breath was focused and intentional. Whenever I would start to drift off I would jolt back awake gasping for air. Drifting somewhere between dreams and visions time crawled by. I looked at my watch; it was six p.m., only two hours had passed. The distant howl of the storm convinced me I was better off right where I was. It was going to be a long night. The snow cave was a foot short causing me to constantly bend my knees driving my aching legs insane. I tried to lose myself in my imagination. Let my thoughts carry me away. I thought of the Haiti earthquake and those that were trapped in the rubble for days on end. It must have been a thousand times worse than this. I thought of Kate, and wondered what she was doing. I thought about all the things we were looking forward to in life together: our wedding, a place of our own, seeing the world, Cloey. Was it worth it? If I died, buried under six feet of snow in the middle of nowhere, would my life to this point count for anything? Would I have done it differently?

It was 2:00 a.m., ten hours since I closed up the entrance. Condensation began to accumulate on the inside of the emergency blanket and steady drips from the ceiling were a constant reminder of where I was. Complete silence, complete darkness, everything was wet again. I brought a can of peanuts into my bag to pass the time. I ate a single peanut and counted to fifteen, then ate another and counted again. Two hours later the peanuts were gone, it was 4:00 a.m.

Above - Changing into my last two dry long sleeve shirts, and dry jeans for the night. I knew at this point I would most likely survive. All I had to do was wait.

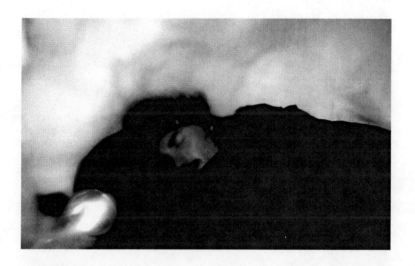

Below - 15 hours later... Attempting to find a happy place for the remaining two hours of darkness. The empty can of peanuts next to my head.

CHAPTER 6
ESCAPE FROM SIX FEET UNDER

Six thirty a.m., my alarm sounded, relieved that the long night was over I unzipped the sleeping bag and crawled to the farthest corner possible and peed one more time. The multiple streams of pee had caused a deep narrow hole edged with yellow that descended even farther down into the snow covered mountain slope. The hush howl of the wind was gone now, whether because of the additional snow blanket or the calm weather, I didn't know. I did know that either way, the transition from snow cave cocoon to running would be freezing so I forced myself to wait another two hours for a few extra rays of sun. After being confined to the snow cave for 16 hours it was finally time. Grabbing one of my running shoes, I smashed at the covered entrance. It was frozen solid from the condensation and additional fallen snow. Luckily I had brought my knife into the cave and began chipping at the ice. A few minutes later the snow and ice gave way and I stuck my head out from the cave like a groundhog. If someone had been watching they probably would have had a heart attack right then as I burst through the freshly fallen blanket of snow.

The two faced mountain had turned its ugly side away reveling blue skies and a morning sun that sparkled over the magical snow covered evergreens. Ruby was half buried in snow and looked like her night was much worse than mine. The clouds had sunk to the valley below leaving me on an island in the sky. It was hard not to feel the pure joy that consumes your essence when life is handed back to you. The island in the sky, I had never seen anything so beautiful.

When I made my way out to the road I was glad to see a snowplow had beaten me to the punch. It had carved out a single lane that would lead me away from my hotel in the mountains. I attempted to run, but my legs were extremely cramped up from being in the confined space for so long. Jeans were not exactly ideal running pants but were still the warmest option. A slow walk was all my body could handle, and even that proved to be too much at times. All of my gear was wet. It was a cold winter day but calm compared to the night before. I knew I had to keep moving, a 45 mile run looked like heaven compared to another night in a snow cave. It was only a matter of time before another storm would come ripping through the mountains. My knee had a pain that felt more like the start of an injury than my average day-to-day aches. I walked all day only attempting to run a few times. My reflective mood on my own life to this point entertained me. I contemplated deep thoughts and unanswerable questions. Step by step persistence moved me to another place and time elapsed.

After about ten miles in the sky, the road began to descend. For twenty miles Ruby and I wound downward the snow slowly began to disappear replaced by a stream that gushed icy water. I rolled into Tres Piedras at about 4:30. It didn't take long to figure out it was a ghost town. There was a fence around the only gas station. I passed through the town that I had strived so hard to reach in less than five minutes. It was another 29 miles to Taos, my only option was to keep going. A couple miles out of town a man in a small truck pulled over. He yelled out the window, "Are you Abe Clark?"

"Yes, how did you know?" I asked.

"You fit the description. I work for the DNR, the state patrol was notified that there might be someone stuck up in the mountains so they drove through before the storm got too bad looking for you. They figured you must have stayed back in Chama or something until the weather got better. They didn't see you on any of the roads. How did you make it this far anyway?" He had an open can of beer in the cup holder between the bucket seats of his small truck.

Not knowing where to start I replied, "Well, it's a long story, but do you think you could give me a ride to the nearest hotel?"

"Well, not like I have any other plans, hop in, you can throw your stroller in the back."

We sped through the treeless valley as fast as the little pick up would take us. "Mind if I smoke?" he asked. "Actually it would be nice if you waited until you dropped me off. Sorry."

He vented about a wide array of topics, swearing with every other word. He continued to vent the entire way before dropping me off at a small hotel that he clamed would be the cheapest in town.

I thanked him multiple times as we unloaded Ruby. Reaching out to shake his hand he said, "I don't mean to be mean but do you have any gas money? I need some beer money while I'm in town."

Slightly caught off guard by his bold statement I turned to Ruby. "O yeah sorry, no problem, you probably saved my life, it's the least I can do." I handed him a ten-dollar bill and he hopped back in the truck.

"Good luck on the rest of your run." He shouted as he sped off.

I would figure out everything in the morning. For now sleep was my sole priority. It had been a long two days.

The next morning I talked to my relieved mother. She

had sensed I was in danger and prayed the entire night for me while I was in the snow cave. The closest church turned out to be four miles away located on the other side of town. They have changed a lot of the area codes around here so none of my phone numbers were working and I had no idea what time the service started. I walked in halfway through the service; every one turned around and looked. There were only about 15 people in the congregation. I slid into the back row completing my failed attempt to go unnoticed. A man came from the sound booth and handed me a bulletin. I quickly scanned the pages before my eyes stopped on the words, "Pot luck dinner after today's service!" Perfect!

We sat around a table in the back room eating and talking. Everyone was very kind, and I can honestly say I left knowing the whole church. The Tedesco's welcomed me into their home for the night. On the way back to their house we drove past Taos Pueblo, which was on old Pueblo village that had been there for over 1,000 years. The Indians that lived there still practiced all of the traditional ways of life. It was like taking a step back in time. The Tedesco's had five dogs, two cats and one baby girl. It felt good to have a couch to lie on and to take a classic Sunday afternoon nap.

Jamie Tedesco drove me out to where I pathetically stumbled to two days before. I was out of the mountains even though the elevation was still above 7,000 feet. I had about 26 miles left to get back into Taos and the gradual down ward slope of the valley made it a very pleasant day. Over the past few days I've observed that Taos has a strange assortment of extremists. The first being the Pueblo Indians that still live exactly how they did 1,000 years ago. Now to my left lay an earth ship community of 90 homes. Alternative housing ideas have always fascinated me and I have read many articles about these earth ships. I was thrilled to see they had one open to the public and I gladly paid the $5 door fee. Earthships are made out of recycled materials such as tires and bottles and are completely off the grid, meaning they produced their own water, electricity and heat. If I ever build a house I would love to use many of their techniques which allowed them no monthly bills.

Down the road a couple of miles the hippie culture that I have heard so much about finally showed itself. Next to the 800 foot deep Rio Grande Gorge were hippies, Indians and people who lived off the grid, attempting to capitalize on the unaware helpless tourists. People go to fairs, festivals and on vacation with money and they HAVE to spend what they saved. I believe this is common knowledge among vendors and they always do their best to get it out of you. One man in particular yelled out his well thought tag line, "Rocks, sage and pipes! Some things you need: some things you don't." His product didn't interest me in the least, but his mini yellow bus with inspirational quotes of being free and a smoke stack coming out of the side window did. I thought his way of alternative housing rivaled the Earth Ships maybe not as efficient but definitely as creative and much more portable.

The sunny warm weather left everyone in a good mood as I made my way down the line. Most of the crafts were extremely simple such as dried out cactus sticks tied together to make a cross or leather pouches that looked like what I made when I was a ten-year-old playing mountain man. It was a great day that was made even better by how good my legs felt after yesterday's day off.

Day 44 (Angel Fire) 23 miles
"I packed up my things for the 44th morning in a row – it felt different. It no longer feels like a trip or even a long adventure. It feels like my way of life. I ran west on Highway 64 today towards Angel Fire. Angel Fire is basically a ski resort town at the foot of a large ski hill in the mountains. The winding road was up hill the entire 23 miles. Strong winds bounced around the canyon

like a 1960's pinball machine. One minute it would be pushing Ruby up the hill without my assistance and the next it would be blowing my hat off sending me scampering after it in the opposite direction. After I climbed as high as the mountain would allow, it shot me down into what I would describe as a valley in the mountains. Although in the mountain valley, Angel Fire was still at 8,000 feet elevation. I am staying at the Elk lodge, which is about a mile off my route. The extra two miles turned out to be well worth it. I am the only one staying in this fifteen-room lodge due to the skiing season coming to an end. The manager made me a big pasta dinner and we had a good discussion about her children's college plans. The big stone fire place with an elk mount hung over it and bear skin rug below it make this place a classic mountain lodge. I have never seen larger stone floor tiles in my life."

I entered my downward descent after a slow first few miles, due to the distracting beauty of Eagle's Nest Lake that lay in the snow-covered valley. Cimarron Canyon wound its way down out of the Rockies towards the Great Plains. The curvy road reflected the bends of Clear Creek that ran parallel to the road. Tall trees took advantage of the water source, packing the narrow canyon from wall to wall. The high piles of snow have turned to patches and then disappeared all together. It was a perfect day with ideal scenery only dampened by an occasional car that would speed around the tight blind corners and forced me off the road.

The welcome to Cimarron sign read, "Where the Rockies meet the Plains." I had anticipated being thrilled to be done with the grueling Rocky Mountains, but instead felt almost sad to see them fade. Over the last two weeks my sweat had bonded with the rugged landscape. Like a good friend, I had seen them at their worst and at their best. They had tested my limits and rewarded me with their beauty. Friedrich Nietzsche, a philosopher, wrote in his book *Twilight of Idols* the popular quote, "That which does not kill us only makes us stronger." This rang true for my journey through the mountains. There was no doubt in my mind that the Rockies had made me strong, because I sure wasn't dead.

Above - Crossing the snow covered landscape of Carson National Forest.

Below - passing both the 1,000 mile mark and the Rockies.

CHAPTER 7
LIFE ON THE PLAINS

The Historic St. James Motel in Cimarron housed many famous people through out the years. the most notable being Wild Bill and Annie Oakley during one of their Wild West show tours. There were a lot of quirky aspects to the place, but what stood out to me were the mounted long horn cows in the continental breakfast room. The waitress most likely found me a bit strange because I couldn't stop laughing. I had the funny thought in my head of people in Wisconsin mounting dairy cows. I'm sure I was just hard up for entertainment. After breakfast I headed out to conquer my new and strange surroundings. The gradual rolling hills went on endlessly with little to see. Ruby rolled easily which always makes the day a little brighter.

I came over a hill to see a large herd of buffalo on my left. They dotted the open fields along a winding creek that flowed from the mountains. Intrigued by the uncommon sight, I parked Ruby and grabbed my camera. I crept through the small trees along the creek for a closer look. A male, of prime physical condition, stood twenty yards from the thicket where I stood. He ripped the grass from the earth, with his huge teeth grinding it down to fine mulch. At least 50 others stood in my view grazing the vast field in plain view. Looking through the viewfinder of my camera, I snapped numerous pictures of the enormous beasts. The Rocky Mountains still towered behind them in the distance adding to the powerful scene.

A twig broke from one of the small trees as I repositioned myself; the large male closest to me raised his gigantic head. His brown coarse hair appeared to be a mix between an African lion and an 1800's mountain man. His ice-cold stare burned a hole right through my pounding heart. His comparatively twig like leg stomped at the ground. A few others followed his lead and looked up to see what the commotion was. I stood helpless and motionless in the thicket. Not a muscle in my body moved other than my increasing pulse. They must have had all the time in the world to waste because it felt as if our frozen gaze remained locked for an eternity. I stood determined not to move until their attention returned to the oceans of grass. The last thing I wanted was a stampede of buffalo after me. The moments passed, and his head went down. My body went into a slow reverse motion placing each foot in the exact spot it had come from.

A song ran through my mind, "Oklahoma! Where the wind comes sweeping down the plains." I was not in Oklahoma yet but I was sure getting close, and I think the wind was getting a running start before it hit the state line. A later weather report recorded wind gusts of up to 50 mph. At times I would get going so fast I would reach a full sprint. Ruby tugged on my arm like a hound dog does when it smells a coon. It felt like I was running downhill all day long and my constant attempt to slow down made my legs burn. I ran the first 15 miles in two hours and sat down exhausted from my failed efforts of keeping Ruby under control. I was really out in no man's land now. It felt like I was running on a giant grass treadmill. The road disappeared into the horizon and a slight change in a field's brown hue was often the most exciting part of the landscape. There were a lot of antelope around that broke up the redundancy of grass. When they spotted me, they would bunch up and flow through the waves of prairie grass like a school of minnows.

Day 51 (Clayton) 32 miles

" The wind has not backed down and blows with even more intensity than yesterday. For the first time in 1,100 miles, I charged up my IPod in hopes the music would drown out the wind. I have to be honest, the little things are starting to get to me and the wind is pushing me over the edge. Ruby is constantly out of control making a steady pace impossible. I'm usually not thrilled to see a hill, but on this particular day a hill meant that Ruby would slow her relentless high speed down to a pace I could handle. At thirty-two miles I reached Clayton and looked at my IPod, it read 127 songs. The long stretch on the giant windy treadmill was over, at least for today."

The wind had shifted to the north and had a cold feel as it passed by. It blew so hard it would pick up Ruby's front tire and spin her around. It felt like I was trapped in the wind, as if I was submerged in water, it filled the space around me. The winds presence was suffocating and its movement constant and annoying. I screamed at nothing. "Please God, let me go," my words lost to the wide open. Two miles from the Oklahoma State line I sat down on the guardrail. I had enough for the day. The wild wind was just too much to handle. However, like so many times before, time passed and eventually I rose and began putting one foot in front of the other. Two miles later I left the beautiful but challenging State of New Mexico and jogged into Oklahoma. I never looked back, I think I was afraid I would see my frostbitten toes blow by in tumbleweed. New Mexico was done, and I didn't shed a single tear, not that anyone would have noticed due to the wind.

The north wind was still blowing the next morning but died down quickly and it turned out to be a very nice day. My legs had become sore over the last couple of days due to the strange running pace the wind had forced. Now that it was calm I could finally think clearly. You can see for miles and miles out here making the approach into the next town painfully slow. I would think to myself, "Yes I made it!" Then ten miles later I would plod into town.

Day 53 (Boise City) 30 miles

"Last night I spoke at an adult Bible study and met the pastor whom I would be staying with once I got to Boise City. The days are rolling by uneventful, much to my relief, and I made it to town by late afternoon. With the gentle breeze causing the oceans of grass to ripple there is a unique freedom out in the plains that was much different than the mountains. It felt similar to flying in an airplane when you come up through the clouds and find yourself sitting in a whole other world where there is nothing but a soft white floor and endless blue skies. Pastor Ordean Nelson of the First Baptist Church of Boise City Oklahoma came out to pick me up and take me back to his house. The next-door neighbors came over for dinner and I met one of the funniest people I have ever come across, Todd."

Todd Toon was one of those people that made life simple and really brought out the true joy of being alive. He seemed to have it narrowed down to good friends and good food. Some people just have that pure excitement for life that bubbles over and makes everyone around them happy. Todd was 38 and diagnosed with Down Syndrome, that however has never discouraged him in his pursuit of the ladies. As far as he was concerned anyone is fair game to flirt with including the Pastor's wife, Pam!

Over dinner Pastor Ordean, headed up the conversation, "I believe every small town has something to be proud of, something they are known for or rally behind." I must admit the two stories he mentioned both made me a bit depressed. Todd focused on his food as he went on. "Boise City first claim to fame is being one of the only cities to ever get bombed by our own military!"

He waited for my reaction; I widened my eyes as I said, "Really, how did that happen?"

Happy to see my interest he went on, "Well, of course it was an accident; the pilot thought the ring of lights outside the courthouse was his practice target. I don't think he was quite ready for battle, but on the bright side he did hit his mark." We all laughed as the dinner conversation turned toward the other thing Boise City was known for, the Dust Bowl. From an outside perspective it seemed to be another depressing memory that didn't seem any better than the last and I wondered if the town had a bit of a complex.

The next morning I ran to Keyes. It was another beautiful day and the 16 mile run only took a couple hours. I ran fast probably because Pastor Ordean and his wife Pam promised to take me horseback riding on their friends 2,000-acre ranch. After dinner with the Toons we headed out to find the horses. Todd had about 80 pounds to lose before Pam would let him try riding her horse, never the less he knew he would get there some day. He was still excited, he was getting close to his goal. We drove down dirt roads for 40 minutes before we reached the ranch. The horses were down in the valley next to the watering hole. With the sun getting low in the sky we saddled up. There were no trails just the wide-open fields to ride across. The only thing you had to watch out for was prairie dog holes and rattlesnakes. Pam's horse nearly stepped on a rattler causing the horse to go into a bit of frenzy. We rode until sunset and even though I still had my Asic running shoes on, I felt like a real cowboy.

As I made my way to another small town on another day, a large pillar of smoke rose up on the north side of the road. It looked strange, the smoke was growing fast but I didn't see any flames. The wall of gray smoke grew to about a quarter mile in length and stretched equally high into the blue sky. I kept running and never found out more until later that day. A grass fire had somehow started and burned a large chunk of land. The flames were only a hundred yards from the road but seemed to be under control by the several trucks that had appeared on the scene. They were lucky it was a relatively calm day; otherwise, with the high winds, grass fires are very hard to contain. Lea Lavielle picked me up with her three children and explained that they would mow or plow ditches and drench them with water to cut the fires off, but if it was too windy, the fire would jump the ditch. The run was short and sweet, and I was happy to have another bed for the night.

Day 56 (Rolla) 16 miles

"Michael and Lea Lavielle have three children Hadin (seven), Abbie (six), and Dominick (three). They all made me feel extremely welcomed. We went to their church in the morning and I was able to share a little bit about the mission during missionary moments. Pastor Terrell Giddens came over for lunch at the Lavielle's after church. Around 1:30 I headed out for an afternoon run and was glad to leave Ruby behind for the day. Lea would come out to Rolla and pick me up when I called. I think Ruby and I needed some time apart. I immediately crossed into Kansas and between the good weather and with my cheerful mood it was the makings of a great run. Today as I jogging down the road I thought about how much I loved to run. After 1,200 miles in two months you may think I would start to get sick of it, at times I was. However, today it was quite the contrary, the same excitement of heading down the open road was still as strong as it was on the very first day. I feel very blessed to be able to do what I love so much."

Lea brought me out in the morning to Rolla and planned to pick me up again in Hugoton. Lazy old Ruby took another day off and left the running up to me. I only carried a fuel belt that held 20 ounces of water, my cell phone and twenty dollars. I usually drink more water but since I was carrying it, I tried to scrape by with the bare minimum. It was another nice day to run and my pace was quicker for a couple of reasons: first and foremost, I wasn't pushing Ruby, I didn't have my camera to distract me, and there were no food breaks. I have noticed now that I'm out of the mountains that I can plow through a run in a condensed fashion rather than it being an all day event.

The farther I get into Kansas, the more green I start to see. A few different kinds of spring flowers are starting to bloom painting the ditches of the road with yellow and purple. When Lea came out to Hugoton she brought all of the children and we went to the Sonic drive thru. The children were excited to get the toy that came along with their meal. I shared their excitement, but must admit I was really looking forward to the burger vs. the toy. It was fun that the children were starting to feel comfortable around me and were telling me their many stories. After dinner Michael took me over to the shop to fix my computer cord and showed me all of the antique cars he has worked on. There were two sheds full of antique cars. One shed completely housed all of his finished cars, the other full of future projects. I think my favorite was the 1923 Ford Model-T Doctor's style.

I had been at the Lavielles for three nights and decided

to stay a fourth by taking a day off. It was really nice to be in the same bed for more than one or two nights. The three children had come out from behind their reserved stranger shields making it a very fun and eventful day off. Our day passed by catching ladybugs, jumping on the trampoline and Abe propelled car rides up and down the dirt mound. In the afternoon they did their home school obligations and I did some work on the computer. I have to admit that part of the reason I wanted to take a day off was because the Toons were coming over for dinner. The Midwest hospitality seems to have peaked here in Kansas. It seems that the small towns out here never eat a meal alone. There is always some type of guest or neighbor over to share in the food and conversation. At first I thought it was just my presence but after sticking around awhile I realized it was happening with or without me. The day before I got to the Levielles they had nine adults over and 13 children. It seemed like it was just their way of social entertainment. It was fun to see people who I had a memory with even if it was only a few days ago. After dinner we all walked over to the shop to take a look at the antique cars. Todd tried to convince me to let him run the rest of the way with me. However, the two-block walk and the potential girlfriend at the grocery store made us both rethink the idea. It was wonderful being in the same place for a couple of days and building relationships past the point of "Hello, I'm Abe."

We drove out to Hugoton and pulled into the Donut Express where I had left off running. The Donut Express was adjacent to a school, which caught my eye because of its blue rubberized track that lay outside its front doors. I left Wisconsin in the thick of winter and now it was already track season! The length of the journey was starting to show with the change of seasons. I hugged Lea and each one of the children thanking them repeatedly and jogged off. Once again the day off left my legs feeling fresh and ready to run. The 23-mile run came to an end at an intersection where a young reporter was scheduled to pick me up for an interview and bring me to a hotel. The run went by quickly. My only stop was a quick afternoon shower in one of the irrigation sprinklers. I got to the intersection about 45 minutes early and wheeled Ruby down into the ditch to relax. I looked at the map as I laid in Ruby's shade. It was a beautiful day out and I enjoyed the warmth of a fresh summer breeze. A bit of commotion on the road caused me to look up and see the county sheriff pulled over behind another car. I watched the scene unfold and realized they where both interested in me. The sheriff was going to see what I was up to laying in the ditch and in the other car was my contact, Laura. She had red hair that almost matched the red VW bug she drove. She was a recent journalism graduate from a college in North Carolina and although Liberal, Kansas was not her dream location, she accepted the opportunity and was making the most of it.

She showed me all around town. The most appealing attraction was the home of Dorothy from the movie, "The Wizard of Oz." Every year they have a celebration called The Oz festival, which, among other things, includes a Dorothy look-a-like contest. The other intriguing event that rivals Oz Fest is a quarter mile women's only race in which contestants run with a frying pan flipping a pancake along the way. Only Liberal citizens are allowed to enter because the town has a heated rivalry with their sister town in England. The neighboring towns' contestants run the race at the same time and than compare results to determine the champion. I was amazed to later see on the Southwest Times office wall that last years winning time was sixty-three seconds!

Day 60 (Mead) 38 miles
"Laura took me back to the intersection this morning and got a few pictures of me running before she went back to the office to work on the story. There is a town in-between Liberal and Mead but since I had an early start, I skipped it. My legs were a little tight, but felt

strong and confident with each stride."

After the first ten miles I started to get into a nice groove and put my watch on the passing mile markers: 7:43, 7:22, 7:05, 7:00. I started to get a little carried away. Running seems to be the cure for many things. Today I was getting some personal frustration and anger out through the run. I pushed Ruby far ahead and would chase her down again and again. The mile makers were always counting: 58, 59, 60, 61, 62, 63. I never knew how high they would go or when they would start over but they were always there reminding me the next mile would be the exact same distance as the last. My mile times continued to drop: 7:10, 6:25, 6:12. A final shove of frustration sent Ruby 20 meters into a grass field and ended my rampage. I collapsed in the ditch. My chest rising and falling, my legs and arms sprawled out over the green Kansas grass. Today, I hated being alone. People dropped in and out of my life like a passing leaf flowing in a calm but steady current. Would this journey ever end? Would this road I was on ever lead home? I stared up into the blue sky and watched the clouds pass. I wondered if they were lonely up there. I wondered if they to were looking for home, an end. Or did they just float on by riding the transparent breeze. I got back up to finish out the remaining 21 miles.

At some point during the 21 miles I could see the highway take a large arching turn and a dirt road that took a more direct route. The dirt road turned out to be too much of a temptation, I couldn't resist. This route would gave me a chance to get away from the semi trucks that seemed to carry the wind in their trailers. My forward motion literally stopped by the power of their gusts, not to mention the manure bits that would fly out at me. Most short cuts end up backfiring and this one was no different. It turns out that Kansas has a lot of little thorns on their dirt roads that cause local cyclists to go through bike tubes like crazy. Ruby somehow got herself into a mess of them and came down with a flat tire. After examining the tire, I made a pathetic effort to pump the tire back up and continue. The effort didn't even come close to working. I sat in the dirt and took the tire off and pulled out the tube. There were at least three holes in the over used tube making the patch job three times as long. After pumping the tube up for a second time and it losing air equally as fast as the fist time, I decided two wheels would just have to do for now. The awkward position got to be too much, and I knew I had to try to patch it up again. After placing a fourth patch on the tube and pumping the tire back up for the third time I was thrilled to see it held air. My excitement was short-lived when it was flat a mile later. I had run out of water and was starting to feel dehydrated. My only choice was to push Ruby through the dirt with a flat tire. I started to stumble and felt as if I was going to black out. I knew I had to get to some water. A few long miles later an old farmhouse rose from the dirt fields like an oasis in the desert. I walked right up to the house and dunked my head under the hose then filled my water bottles. I could feel the cold fresh water trickle down my insides and bring my body back to life. Ruby and I limped the last ten miles into town listening only to the thump, thump, thump of Ruby's left tire.

Jerry and his family were from Wisconsin and insisted on feeding me all sorts of classic Wisconsin food which included cheese, crackers, Point Root Beer, deer venison and five gallon buckets of honey. Jerry worked at the local Christian radio station so we went over in the morning to do an interview. It was raining out, and the weather showed it would clear up in the afternoon so I waited around at the station. Jerry's oldest son was in the Army and currently experiencing his second tour in Afghanistan. He was talking to Jerry on Skype, being in the room at the time I voluntarily dropped in on the conversation. It's incredibly hard for Americans to relate to these men and women; the news can't explain the reality of these soldiers day to day life. After listening to a father talk with his son who is in constant danger and thousands of miles away, I gained a new perspective for the situation. War makes me sick but for some strange unexplainable reason it is something that is always there. Jerry told me his son's worst experiences have been picking up body parts after a suicide or car bombing. The thought had never crossed my mind.

The next day I was invited to have dinner with a family who had seven children that ranged from the ages of one to sixteen years old so I eagerly made my way down the straight Kansas road. Now that I have a few states behind me I feel I can confidently say that Kansas's roads have great shoulders, which makes for fast mindless traveling. A herd of cows took a break from their grazing to run alongside Ruby and I. Lately my legs have been stiff and sore because of the redundant muscle use caused by the nonexistent hills. Minneola was another small town. I meet a kind man named Lukas at his home, settled in, took a shower and we headed over to the Borchards for dinner.

I have been around large families before and growing up with four siblings of my own, I have had a taste of what it is like. On my run over I thought of the movie *Seven Brides For Seven Brothers* and wondered if it would be anything like that. Upon my arrival I was welcomed into an orderly household filled with a few of the politest children I have ever met. It was truly a joy to sit at the large table and listen to them share their exciting stories, one by one of course. After dinner the older ones played their favorite piano song for me including Rebeccah who had a broken arm. Apparently three fingers on her right arm were enough to rock her song. We put in a movie and all watched Don Knotts star in some crazy version of *The Love Bug*. I really enjoyed meeting the Borchards. It wasn't until later I realized all of their names ended in ah: Johannah, Micaelah, Jedidiah, Rebeccah, Hezekiah, Ariannah, Adoniiah.

My destination was Bucklin, another small town along Highway 54. It happened to be prom night there, which apparently is quite the shindig in Bucklin. The run seemed to take forever, and I couldn't convince my lazy legs to corporate.

Once again I could see the town in front of me, but it was still at least six miles away. One thing about Oklahoma and western Kansas is there are no trees. In other words for a long distance runner, your bathroom door is open for all eyes. There have been a number of times where people passing by saw me crouched down on my imaginary toilet. After our eyes meet there is usually nothing to do but wave. Faced with the same situation today, I was standing on side of the road in a bit of an awkward stance. As I waited for a gap in the cars, the next car slowed down to half its speed and stuck a video camera out the window as it passed. Once again, caught in the act, there was nothing to do but wave. Like so many other times before in the last two months, a steady dose of persistence brought me to my next destination. I finally arrived at the Bucklin City limits sign. I called the Pastor and he said they reserved a room for me at the local motel, because every one in town was busy with prom.

Before church Pastor, came to pick me up, and we headed over to the annual Biker's Breakfast. He said they had been inviting him for years but he had never gotten around to it. It turned out to be a different type of bikers' crowd from what I was expecting. I must admit I had some trouble finding topics to discuss with them. I was, however, very grateful Pastor had taken the time for me. Between staying up until 3 am at prom and having to preach in a couple hours, I'm assuming the last thing he wanted to do was go to a biker's breakfast. He dropped me off at the hotel and I packed up my things. I walked over to his church for service. The Pastor called me up to share for a few minutes. I talked about how I had run 1,400 miles and about the mission of Living Water International.

The run to Greensburg seemed long just like the day before. Along the way some guys were riding around on their four wheelers with gas blowtorches lighting the whole prairie on fire. It looked a little fishy to me, but I guess they knew what they were doing. The flames burned through acres in the short amount of time I was there to observe. It has been raining off and on throughout the last couple of days and the spring flowers have made their homes in the ditches along the road. For the last 100 miles I have heard people talking about Greensburg. Back in 2007 there was a huge tornado that took out 95% of the town. The government adopted the town because of its name and helped rebuild everything to be energy efficient. Before all of this took place, the pride and joy of Greensburg was their largest in the world hand dug well. Now everyone talks about the solar panels and wind turbines. I was amazed by how strange it looked. Every other house was either brand new or completely trashed. Every single tree appeared to be half of its original height and there was still a road sign folded in half from the high winds. Grass peeked through the cracks of cement where gas stations and buildings used to be.

Day 64 (Pratt) 31 miles

"My housing situation was a bit confusing last night. After walking around town for a while, I decided to wait at the Methodist church for my contact that was arriving home from Dodge City. It turned out he was the pastor of the church I was waiting at. It was a large church, the largest I had seen in a long time. As he called it, I was not the only transient in the area. There were two others staying in the basement rooms. As he showed me around he opened a storage room to get me an air mattress and a half clothed young man with dread locks sat up from a similar air mattress. He muttered something, but I couldn't understand him. When I awoke in the morning, he was gone. I wandered up to the kitchen and made some macaroni and cheese in the microwave. There was a man doing some electrical work and he told me a little bit about the tornado that had come through. The conversation lead from one thing to another and eventually he shared with me his jail ministry that he and his wife had started. When the conversation had run its course, I told him it was nice to meet him, washed out my bowl and left."

The run actually went well today for some reason. It was on the longer side but I chipped off five or six mile stretches like clock work. The closer I got to Pratt, the more trees I saw. There seemed to be an increasing number of streams and rivers which the trees stay close to, binding their roots to the water in order to survive. A couple of hotel billboards gave me some numbers to call, and I fished around for the best price. I walked into the first one I came across. She said, "Looks like you've been traveling."

I replied, "Some call it that." The price of four walls and a roof had been cheap as of late but this one cost 50 dollars. There are few things better than curling up in a big bed after a 30 mile run. One of them is curling up in a big bed with a large Pizza Hut Supreme Stuffed Crust.

My legs were pretty shot, and I took every kind of recovery drink, capsule and gel I had. I even took some time to stretch, which was an activity that usually never happened. At 11:30 a.m. the front desk called and asked if I knew that check out was at 11:00 a.m. I pretended to be oblivious to the situation, but was well aware of the rule. Not only was the fact stated on my door but it also seemed to be the standard check out time for every hotel I've stayed at since the Pacific Ocean. They did, however, grant me my only wish, which was to stay a half-hour longer. Then I was kicked out onto the all too familiar open road. My running effort was sub par, and I just couldn't seem to get my legs going. I felt so lethargic. The highway traffic picked up, and I began to wonder if the line of telephone poles a half mile north had some kind of frontage road next to it and followed the same route. I crossed a field of grass and came up on to the calm road that had peaked my curiosity from afar.

The road was gravel, but the difficult way that Ruby rolled on it discouraged me from running further.

It didn't take me long to figure out there was no way I was going to make the 30 plus mile run to Kingman today. When that thought hit me, I stopped and couldn't get passed it. There was a big square cement culvert that had just been put in and it looked like a great place to sleep. It was no later than 3:00 p.m., but I rolled Ruby into the four-foot high cement tunnel and set up my room for the night. I fell asleep for a while, then just lay there listening to my new surroundings that were enhanced by the echoing tunnel. It's not like I was exhausted or anything, but my legs had no interest in running today. It was the first time I had given up on trying to make a checkpoint to sleep in the ditch. The gravel road ran next to the railroad tracks, and before I made myself a fine meal of pita bread and raisins, I walked over to the tracks. I placed my only sixteen cents down on the rails.

The culvert was new and I loved it. I thought about the renovations I would make if I decided to stay there for the summer. I tried to predict the movements of the waving branches outside but never could. I listened to the funny echoes in the tunnel and tried to roll rocks down the whole thing and was excited when they dropped out the other side. I moved my bed around a couple of times to find the best sleeping spot. I read the labels on the cans I had and read a few stories in the Bible Lucas had given me. I watched the overcast sky turn from light to dark and listened to the distant chug of an oil pump motor. I fell asleep.

Above - Packing up my things after spending the night in a culvert.

Below - Giving Abby a piggy back ride to her dad's Antique Restoration Shop in Elkhart, Kansas

Above - Over looking the Black Mountains on the original section of Route 66.

Below - A torched car with hundreds of bullet holes in it lays abandoned back in the desert hills.

CHAPTER 8
MIRACLES AND DEATH

Day 66 (Kingman) 22 miles
My diet is constantly changing depending on what kind of food people feed me or what is available.

Monday – pita bread and peanut butter, bowl of macaroni and cheese, peanuts and raisins throughout the day, box of granola bars.

Tuesday – pita bread, peanuts, raisins, 3/4 large pizza hut supreme

Wednesday – pita bread, rest of pizza, six pecan cinnamon rolls, peanuts, GU Energy gel, triple burger, French fries

I slept well in the tunnel and found it to be warm throughout the night. Ever since the snow cave episode I've been hesitant to sleep outside, worried I might freeze to death. However, the lower elevation and approaching summer proved my thoughts to be useless worries. I packed up and pushed Ruby back onto the gravel road. I was certain that my 16 cents had not been flattened because I didn't wake up in the middle of the night thinking I was going to get run over by a train. Although I was disappointed, it was probably for the best, I had gotten a good night's sleep and my sixteen cents could now be invested more wisely.

A church in Kingman bought me a motel room for the night and I gladly accepted. The next morning I woke up to pouring rain and a thunderstorm. I checked the weather station, which eagerly shared the news of scattered thunderstorms all day. The weather had caused a power outage throughout the little motel leaving the clocks blinking red. Around the same time, I was kicked out of my motel room thus back on the open road. I noticed, I seem to run better when the weather is slightly extreme. It puts me in a focused and intense state of mind.

Gray lines streaked down from black clouds and every once in awhile one would pass over Ruby and I. I put the clear plastic rain shield over Ruby to keep things dry. Luckily I was blessed with waterproof skin and did pretty well. The highway traffic picked up as I approached Wichita, so I stayed on the side roads that ran parallel. Someone told me 65% of the roads in Kansas are gravel because they are cheaper and require less maintenance. On the down side they cause a lot of flat tires, which I found out first hand last week. Around 6:00 p.m., my side road ended. Between this, the increasing rain and my lack of a map I decided to call my contact. I knew it would be challenging to get through Wichita without my GPS that was lost or stolen at some point last week. There were only a few roads that had bridges over the river so I couldn't veer off my route. Other than that I basically just had to continue northeast towards Bel Aire.

About an hour or so into my run, my eye caught a carnival of some sort going on in a park off the road a bit. There seemed to be two different events. On one side of the road it looked like a Renaissance Fair. It made me feel slightly better that I was not the only guy in Kansas who wore tights and carried a big knife. The other event was a 5k Stride and Ride for MD (Muscular Dystrophy). I usually try to participate in local events but a 5k was not really what I had in mind for a break. On the up side, I already had a seven-mile warm up in. I started to walk back to my route when I decided to run the race and turned around. As I was signing up a lady that was walking around getting "We won't sue you if we die" signatures and smiling slightly bigger than her face would allow, asked me if I was pushing Ruby for practice. I don't think we understood one another and the moment passed without me realizing she meant practicing for pushing a person with MD. Andy the Armadillo and Bethany Gerber, Miss Kansas, were there adding to the upbeat atmosphere. I was able to meet a girl recently diagnosed with MD who was also a huge Packer fan. Packer fans are a little hard to come by in these parts, and I felt like a fair weather fan when she asked me how I thought the draft went. I didn't even realize it had taken place, much less know who the Packers picked. Bethany started us off, and I quickly made my way to the front and was soon running alone once again. The out of the way miles passed quickly and instead of hanging around for the post run festivities, I continued through the finish line and straight into my 23 mile cool down. Eventually I made my way through Wichita and back to the straight brainless road I had become accustomed to in Kansas.

The gentle breeze and clear blue skies took over the thunderstorms that had been lingering overhead the last few days. My mind was excited to run today and my legs for once agreed. I didn't have housing in the next couple towns so there was no limit on how short or how far the days run had to be. I set an impossible goal of making it to Yates Center, half a century away. I knew I would have to keep a good pace going all day long if I would make it before dark. For some reason I just wanted to see how far I could run. I never found out.

I arrived in Eureka around 12:30 and stopped at the Sonic for some lunch. At Sonic there is no indoor seating. You're expected to drive your car up to a station and order from your car and the waitress will bring your food out to you. Ruby and I stood in the parking slot between two heavy-duty pickups. It looked like a hot spot for the area cattle ranchers. Moments like that used to be awkward, but now they pass by me without much thought. The waitress brought me a chicken sandwich, a large PowerAde and an M&M ice cream. The ice cream came highly recommended by the Lavielles who I had stayed with in Elkhart.

After my half hour break, I continued down the road. Either lost or stolen, I no longer had a GPS or map at this point but it didn't really matter. All roads in Kansas seemed to head straight north, south or east, west. It had become my instinct to run east. If I would turn around for whatever reason, the world seemed backwards. My body worked flawlessly and every part of me knew the repetitive task so well it came with no effort or thought. If I thought about my repetitive motion, I could feel every individual muscle and where it connected to the bone. Most of them were sore, but I liked them that way. It was how we communicated. It was a pain that I was familiar with, a pain that I could control, a pain that my body now accepted and at times welcomed.

I made it to Yates Center just before 7:00 p.m. I wanted to run until I fell over exhausted but the moment never came. Over ten hours of running later, my legs felt the same as when I had started. I came to a hotel in Yates Center. Half of me longed to keep running into the night to fulfill the desire to find my body's breaking point. It is a longing I need to know every now and then and lately it has been hard to find. The other half of me knew the road would be there in the morning, and perhaps then I would find my breaking point.

In the morning I piddled around the hotel room like I usually do mixing drinks, making peanut-butter and honey sandwiches, stretching, looking at my route and the weather. My 11:00 a.m. checkout deadline always seems too come too fast. My legs felt better than I expected after yesterday's 52 miles. They finally kicked me out of the hotel around 12:30 p.m. and I took to the road. My legs were a little less enthusiastic about today's run, but they ran just the same. The wide shoulder had narrowed to a sliver forcing me to play chicken with every car that approached. I usually won but every so often there was a semi truck in the opposite lane, and I could tell the cars would rather hit me than an approaching semi. Repetitively being forced off the road into the grass ditch got old fast but there wasn't much I could do about it; if I was going to get hit by a car it was going to be from the front. Right side of the road running always freaks me out and I obsessively look over my shoulder at every car that approaches. Nevertheless it was necessary at times to change sides of the road because of the way the road is crowned. Running long distances on a slanted surface messes with your hips and apparently my left outer ankle.

I made it to Moran at twilight and instead of trying to cram in the last three miles of my run before dark I sat around and watched the sunset and turned around and watched the moon rise. It was a bright full moon so instead of being blinded by darkness, my eyes adjusted to the new blue light that shed over the unknown. When I reached for my cell phone to get the final direction to my contact's house it wasn't there. I pushed Ruby through the uneven grassy ditch about a mile back to find my sunset and moonrise observation perch. I thought it might have bounced out of its pocket back there. My paranoid mind ran in circles; losing my cell phone would be devastating. It was my connection to reality.

As usual I ran fast in the dark; there is something calming yet intense about how the darkness takes away all the detail that usually distracts me. I ran on through the echo of my own heartbeat and in the presence of no one except my moon lit shadow. Traffic had died down but the occasional startled driver swerved sharply as his blinding headlights fell over me briefly bringing me back into existence. When the lights left and the sound drifted off down the road, the trees quickly put up their shields of darkness and let me pass unnoticed.

The road rose out of the valley and the woods let me go revealing a church sign that was lit up by a small ground light. I slowed to a walk as I pushed Ruby up the gravel drive way. Lights were on and a large TV lit up the living room. I knocked. A stocky man with a full beard, reddish brown hair, and small kind eyes opened the door. I introduced myself, and he did the same. He could have told me he was a lumberjack or motorcycle enthusiast, and I would have believed him on the spot. However, he was neither and simply went by Pastor Tom.

I quickly launched my sob story of having lost my cell phone and hinted that we should go look for it before the battery ran out. We drove back to the spot where I had pushed Ruby through the ditch and I used Pastor Tom's cell phone to call mine, listening closely for the distant ring tone "I'll fly away." No luck. I ran along the road scanning my eyes for flashing lights and listening. Pastor Tom had stopped to help a young lady change her tire, and I continued looking for my needle in a haystack. I walked and ran along the ditch, listening closely for two miles. I checked the spot I had gone into the woods to use the "restroom." Nothing. Pastor Tom found me disappointed on the side of the road and said we would come back in the morning and look.

My troubles seemed small compared to the young girl Pastor had just helped. Her tire had blown out, and she had no money for a new one. She was sitting on the side of the long Kansas road with no cell phone at all in the dark. The man she was escaping from was trying to force her into prostitution and the miles between him and her did not seem to grow quick enough. Pastor told her to follow us back to the parsonage where we would take care of her. You could tell she didn't trust us, but she had no choice. When we returned to the house, I woofed down everything that was offered. If it weren't for the sweet innocence of Pastor's three year-old granddaughter I don't know what we would have done. There was only one perfect line that could be said in that moment and the little three year-old Eden said it, "I think your hair is pretty." These words were spoken out of complete innocence from a heart and mind pure as an angel. Through her tears she smiled. Her hair was unique, dyed half red leaving the other half dark black. Pastor asked if we could pray with her, and she agreed. There is absolutely nothing more powerful than a prayer to God in your darkest times. I wish we didn't always leave it to those dark moments but it's a comfort to know when the world fades and crumbles around us, he will always be at our side loving us exactly the same. I said good night and headed to my room that was in a separate building. It had been a crazy night that all made complete and perfect sense when I pulled my cell phone out of a pocket I had never put it in before.

I stayed in a building that was used by the church for group meetings and people passing through who needed a place to stay. I stayed around until late morning and went back over to the parsonage to say goodbye and have some cereal. Little Eden was still in her Lighting McQueen slippers and looking cute as ever. The young lady that we picked up the night before with the blown out tire was long gone. After eating a light meal, I was on the road again. My destination was Fort Scott, Kansas, a small town that lay only seven miles from the Missouri border.

The high mileage of the past couple of days left my legs feeling heavy and worthless. I took my time with the day's run finding excuses to take a break, whether it was to take a picture or eat yet another energy bar. Life was full of dull thrills and mundane tasks. I was lost in a world of constant impulses, striving to complete simple tasks. Once I found a safe place, I always felt like staying forever. Yet in the morning, I was always expected to run on and I did, mile after mile, hour after hour, random place after random place. The constant change and day-to-day instability was beginning to have a strange effect on me. I met so many wonderful people but our relationships were often 12 hours or less and consisted of me telling my story and them telling theirs. I felt outside of life. I was passing through, observing so many random families that lived in a random town. I ate what they ate, I did what they did. I listened to their stories, their trials, and their joys. My route was a time line and I passed over it slowly. The further I ran the less I wondered and the more I remembered. As the miles passed I would think about where I had been and whom I had stayed with.

I reached the Fort Scott church just in time for their Wednesday night Royal Ranger program and was able to talk with the young mesmerized boys about the run and Living Water International. When I was done I asked if anyone had any questions and they all raised their hands. Children ask different questions than adults. Things like, "Do you have a gun?" "Why don't you ride Ruby?" We had a good time, and it was a pleasure to meet such fine young men. They will go far.

The following morning I woke up to a phone call from my sister, Rachel. Early Thursday morning, my grandpa had died. He died in his home in Milwaukee, all five of his children by his side. I immediately called my mom, she said she wasn't sure what the rules were for my run but was hoping I could come home and be with her during the funeral. I assured her it was not a big deal, besides I wanted to go, I needed to go.

There are few things that bring life to a complete standstill; unfortunately it is usually a funeral or a catastrophic event. As I attempted to figure out the logistics of the unexpected break I received a phone call from Kate. She was lost in a world with the ghost of me. She was slowly forgetting who I was and was mad at my absence. Planning a wedding with a myth. A being that no longer existed in her physical life. Engaged to nothing more than a story of a man on the run. I didn't know what to say. My life seemed to be crashing down around me, and for what, this endless road east? Was it really more important than everything I had left behind? I lost it, weeping out of control. I stripped off my clothes and got into the narrow stand up shower. Hugging my knees, I rocked back and forth on the floor letting the hot water wash my infinite tears down the drain. Time passed with the evidence of the water turning from hot to ice cold. Shaking out of control, I begged God to give me an escape from my unwavering commitment to the fantasy of the endless road east and the reality of my crumbling personal life.

Above - Riding a horse in Oklahoma.

Above - Running on the giant grass treadmill of the plains

Below - A refreshing afternoon shower in western Kansas

CHAPTER 9

WHEN A LIFE STOPS

I was at a friendly church where I could leave Ruby and hitch a ride to Kansas City. Plane tickets skyrocketed to over $1000 from Kansas City to Milwaukee because there was no room on the plane. The Greyhound ended up being my only other option and became my home for the next twenty hours. Riding the Greyhound is usually an adventure in itself. Half of the people on there have some kind of a crazy story of why they HAD to ride the Greyhound. The other half is full of tourist from other countries taking pictures at every McDonald's and city skyline. The route I was supposed to go on went straight to Chicago and up to Milwaukee, but of course the first bus was a half hour late and the second connecting bus didn't wait for us. Therefore a few other people and myself, with little other options, took the long way to Milwaukee via Minneapolis.

Staring out the window, from my seat on the bus, the terrain passed so quickly it flew by like life itself. It was a pace I was not accustomed to. Most of the time a blur, however, every once and a while it would come to a stop letting places and objects come into focus. I thought of Kate the whole bus ride, how she must be feeling, wondering if I was ruining our chance of a future together. I thought about all the practical things we struggled with in our relationship. Our wedding was only four months away, and we were far from ready. Sure on the outside we had a place reserved and certain things lined up, all little of my own doing. On the inside we were lost, our connection almost forgotten. We clung to the memory of each other. The memory that, at times felt so clear, so vivid yet other times faded and dissolved like tears falling on an freshly inked letter. Time and logistics were pushing us towards a commitment, a vow that my absence made overwhelming. Life had gotten too big for us to deal with. We were living in separate worlds unable to connect. I found comfort in mine yet in constant fear of the inevitable abrupt end that loomed in my future. I feared that I would not be able to make the quick transaction from rambling homeless man to responsible, supportive husband. She, fearing that our love was too young, too unknown to take on such separation followed by such commitment.

The bus driver made the announcement, speaking in a low methodic voice that gave away his desire for a driving break, "In fifteen minutes we will be arriving in Milwaukee, those of you continuing on to Chicago be back on the bus in thirty minutes." I gathered my things together and prepared to exit. The Milwaukee station was new with tall cathedral ceilings and a wall of big glass windows that looked out into the street. I walked in scanning the bustling hub for one face: Kate's face. I almost willed the moment to exist. Time passed, and the crowd dispersed, loading into different cars or buses for the next leg of their journey. I found a seat and set my bag down.

The morning sun poured through the negative space of skyline. A girl entered the light, half running half walking. She was wearing a new dress that fluttered in the gentle city breeze just above her knees. I sat up and looked closer, a truck blurred past, momentarily blocking my observation. Before another came, I saw the girl had stopped at the cross walk and was anxiously looking both ways. Her light brown hair appeared to float with each turn of her head. When I noticed how closely her hair matched her tan summer skin, I knew it was Kate. Summer was always our season. I ran out of the station, we met on the sidewalk, the city spun around our locked eyes and twirling embrace.

The funeral went as well as a funeral can go. I spent the days leading up to it working on a video to play at the visitation and to give as a gift for family and friends that came. Grandpa had thousands of old pictures on his computer. My mom and I probably laughed as much as we cried sorting through them all. I don't think it's possible to fit someone's life into a ten-minute video. The pictures of happy times were only a small taste of the 85 years of experiences and interactions he had lived.

I played my fiddle a lot over the weekend; I know Grandpa would have wanted it that way. He bought my first violin and was often my inspiration for learning new instruments and songs. When I was little and into playing the harmonica, he said if I memorized 100 songs he would give me his double-decker harmonica. I worked hard on it and only made it halfway before he ending up just giving it to me out of sympathy. I think he liked my effort and enthusiasm. I must say there are two things that stuck out to me about the life of my Grandpa. The first brought me to tears. He was finally with his wife who had died 20 years ago. Mom said that before Grandma died she told her to make sure Grandpa found someone to take care of him for the remainder of his life. Grandpa never even entertained the thought of remarrying. I don't remember too much of them together, but after sorting through their home videos, I was taken to another time. It introduced me to a new view of my Grandpa.

The second thing that really stood out to me was the meaninglessness of material possessions and even our own bodies. Grandpa's belongings weren't packed in his suitcase to heaven. What people talked about was his generosity, the way he loved, the way he cared and invested in other people, his passion for music and foreign languages. Electronic devices and new cars will be outdated before you blink. What you leave behind are your stories. Your legacy is your passion for others. Your portfolio to God is your soul. Death is a constant reminder to us to live. Live with passion, love, have goals, purposes and dreams. A quote I love, "You only get one shot at life but if you work it right once is all you need." Take care of your relationship with God and the people around you. It's truly what matters. I came to the conclusion that a funeral may honor the dead, but it is more for those of us who are left behind. We don't know how to deal with the blank space. We understand little and know even less. Those who die are all pioneers of the heavens. I think it takes courage to die, but perhaps it takes more to live. My grandpa lived.

I spent the week in Milwaukee helping out my family. I found a good amount of time to distract Kate from her finals. I can't describe how nice it was for us to reconnect and forget about the run for a couple of days. We had not been doing well apart, and it was healthy for us to have time together to remember the little things we love about one another. It took me a few days to realize the alternative mindset I had adapted from the run. I had lost touch with reality and slowly forgot normal life. I can't believe how crazy the world is when you stop running by it. People are literally running their brains in circles at a 100 miles per hour. I have long lost any personal stress or worries in running across America. Every once in a while, I will meet someone who is completely baffled by my run. It's so funny because I don't really understand how someone can't understand. I run as far as I can and then find a place to sleep. The simplicity of it is maybe what people don't get? They always ask... "What do you do?" That is the world we live in and although I hope to always stay in-touch with simplicity I'm not ready to lose touch with reality. Kate and I looked at apartments, registered at Crate and Barrel, cooked, walked, talked and sat in silence. For now our reality was perfect.

I had the opportunity to meet some awesome students at Woodview Elementary School in Grafton, WI. Kate's mom teaches there and has been reading my daily blogs to her students since day one. Kate and I experienced a warm and excited welcome as we arrived. The students even made signs reading, "GO ABE GO." They showed me a video of the Destination Imagination competition they did on the Haiti earthquake. Part of the skit was "Abe Clark" played by a young student coming out of a box and running around with a pink doll stroller. It was very funny and incredible acting. The students also raised a remarkable $80 by selling the www.water.cc bracelets! This class is a great example of all the little helping hands making this run a success. The water crises will be solved one drop at a time.

Some friends from Fort Wilderness, the Christian camp where Kate and I met, were getting married the day before I flew back to Kansas City. Kate was a bridesmaid and needed to be up there the day before. This gave me some time to drive up to northern Wisconsin where my older brother was building a log cabin. I was happy to see him exactly where I thought he belonged, in the north woods. It was so far north that it snowed the entire weekend. It may have just been a random storm but still, it's May! We talked about the similarities between Marine boot camp and running across America. He happened to be putting a new pump in his well which I thought was interesting since a lot of the money we are raising will go to well rehabilitation. I can see why someone may need a helping hand in fixing a broken well. We pulled the extremely heavy pump up from eighty feet below. Luckily I was not the one in charge of putting it back together.

The wedding was beautiful, and it was uplifting to see a lot of friends from Fort Wilderness. Kara, the bride cried and laughed the whole time making the ceremony pretty funny. I must say I was a bit distracted by the bridesmaid at the end of the line. We had a good time and it was a fun way to end my random week-long Wisconsin stay. Stopping the run for a week was not a big deal. My goal was to run every step of the way across America in 136 days and that goal would remain the same. I would just have to take out all my planned days off and double any days that were less than 20 miles for the remainder of the journey.

CHAPTER 10

RAINING RABBITS IN THE OZARKS

I packed up my belongings and loaded up Ruby, a routine I had grown familiar with. I can't tell if Ruby was sad because I had left her or she was sad because we were on the road again. In any event she had a flat tire. I had one more spare tube and replaced it quickly before heading out into the rain. The forecast showed storms all day so I covered Ruby with her rain slicker and wore light winter gear myself. The high for the day was predicted to peak precisely at 53 degrees. As I crossed a pair of tracks on my way out of town, a train blasted its horn. Scrambling for some loose change to put down on the tracks, my heart rate and adrenaline skyrocketed. I had been looking for the opportunity to try this for some time now, but I didn't expect it to be this exciting. I laid down a quarter and a couple pennies as the crossing gates dropped, and the red lights began to flash. The train flew by, shaking the ground with its power, and sending what appeared to be my quarter flying off the tracks. When the train cleared, I searched for my money, but it was gone. I couldn't find it anywhere. I finally continued down the road, depressed about my poor investment and contemplating how many other curious travelers had laid their money down only to come up empty handed.

The rain was light and not much of a bother. The joints in my legs seemed rusted together and creaked with every stride. I had not run a single step on my eleven days and my legs were well aware of it. I was already starting to regret the extremely challenging schedule I made for myself in attempt to make up for the lost time.

I stopped at a Subway in Nevada City to get a late lunch and looked over the map. My plan was to run 40 miles today to El Dorado Springs. I knew it would be a stretch with how late in the day it was and how tremendously sore my legs had become after the first day back on the road. I couldn't believe it, I had only run 20 miles yesterday, and every muscle ached. Could my legs really get out of shape that fast? The weather forecast came on the Subway radio, and it proved to be the deciding factor in my plans for the rest of the day. Seventy mph winds, golf-ball size hail and tornadoes were to start around 7 p.m.

The forecast did not worry me; all I wanted to do was run 20 more miles through the Ozark Mountains, but... I had Ruby to think of. We checked into the nearest motel.

The storm passed, leaving rain puddles and patchy overcast skies in its wake. There was no doubt my tent hammock would have been swinging steady if I hadn't happened to hear the weather forecast in Subway. My goal was to make it to Collins today, but with the weather cutting my run short, it tacked on an additional 15 miles. With how sore I was the 41 mile jog seem a bit on the impossible side. The clouds parted to reveal a hot afternoon sun. A policeman pulled over in front of me, and I thought, "Oh great what am I doing wrong?" Hwy 54 was a fairly busy road but I never thought it to be illegal to run on.

He pulled up next to me and rolled down his window. He looked me over once or twice before he spoke, then said, "Someone called in and said you were stranded out here with a baby, but you're clearly out here by choice."

I replied, "Yes sir I am running by choice." I wouldn't think a runner to be a foreign sight anywhere but maybe it was in the Ozarks. Who knows?

I have seen some beautiful places in my travels and would not hesitate to rank the mystical Ozark Mountains right up near the top. I have noticed a lot more animals now with the warmer weather and the abundant creeks. Rivers that flow through virtually every valley astound me. The hot afternoon sun left Ruby and I a parting gift of cool evening air and a gentle breeze. I found myself running over an almost unbroken pattern of rolling hills. A coon crept down from his giant oak tree and poked around in a small creek. Up ahead a cautious coyote crept onto the road. Looking both ways he was surprised his sharp ears didn't hear my silent approach.

The area is well known for its caves and caverns, so when I spotted what appeared to be a cave half way up a 30-meter cliff, I stopped. A few scrapes later and an enlightening reassurance that I was not a rock climber, I sat down on the edge of a little cave. It was a magnificent perch that overlooked the road far below; Ruby looked small as she waited patiently in the ditch. Beyond the road was the rhythmic rolling terrain with thick green trees that cover the Ozark hills like a soft blanket. The setting sun kissed the highest point of each and every tree. I would have spent the night there if I didn't have someone driving out to come pick me up. I left the enchanting hideout satisfied and at peace.

At my host home I read though a whole pile of cards that I had received in the mail from Menominee Tribal School. They must have made them in art class; they were all so creative and well done. They had raised $400 at a bake sale for Living Water and watched my progress as I inched across the giant map in the hall. It rained again last night; I am thankful to be indoors. I picked up my run to Collins a bit more rejuvenated than the previous night. Only eight miles later I ran into Collins with stiff muscles and a general lack of energy. I collapsed in the mowed grass of a gas station. While eating a light lunch that consisted of a Gu, an energy bar and a small bag of trail mix, I looked over the map.

It was 20 miles to Hermitage where I was going to share my passions at the weekly Wednesday night church meeting. The only problem was the service started at 7 p.m., which was now only five hours away. My legs felt completely shot and the 20 miles might as well have been a thousand. However, at this point in the run this situation was nothing new, and after a few minutes of self-pity I simply stood to my feet. I knew I could take one more step, so I got up and took a step. It would have been silly to take a break so soon, so I took another. Before I knew it I was back on the road. My next goal was to make it to the top of the large hill that rose in front of me as I made my way out of Collins. I pictured a king size bed at the top with platters of fruit and ice cream all around me. When I got to the top all I found was another dead Missouri armadillo. Nonetheless, I was at the top. Just as the road rose, it lowered as if it were a reflection in a mirror. Gravity took over and broke my forced steps into a small stride and just like that, I was running again and my legs didn't even know what hit them.

Wheat Land was the small town before Hermitage. Ruby and I rolled in at a pace that made the old man sitting on his front porch bored while watching me pass, and he got up and walked inside. A rabbit hopped out of the tall, green grass and bounced up to the white line. Just then out of nowhere, I heard a thump and turned to see the rabbit flying upside down five feet in the air. He proceeded to take several more flips and awkward twists as a white Nissan sped away. He hit the gravel shoulder with a thud, and then frantically attempted to crawl through the gravel car lot he landed in. His adrenaline took him only so far.

I approached slowly, not because I was scared of the rabbit, but because I didn't want to frighten him further. He lay in the gravel traumatized and shaking on the rocky ground. I looked around the big parking lot; he didn't have a chance. I had seen one too many road kill carcasses on my run not to care. I made some room on top of Ruby and picked up the rabbit. Of course there was the argument that I should have killed him to put him out of his misery, but if I were in his shoes I would have much rather died on a grand adventure than like a fool who ran out in front of a car.

I named him Peter Wheatland after the legendary Peter Rabbit who escaped the wrath of Mr. McGregor and after the small town of Wheatland which he was from. His back two legs were clearly broken and hung limp. I finished out the run and put him under Ruby in the church lawn and walked in the church doors at exactly 7 p.m.

Even for me, it was the first time I had ever spoken in church while wearing half split running shorts. I was thrilled that everyone sat during the whole worship service, but it presented a different kind of problem when I was called up to speak. My legs lock themselves in a sitting position. I awkwardly slid out of the pew and did an odd turn that enabled my body to avoid using my quads. My hamstring started to cramp as I began the long journey down the center aisle. I played it off the best I could; however, I'm pretty sure that everyone thought I was crippled. As I staggered to the altar, I imagined it would have been hard to believe that I had really ran all the way from the Pacific Ocean. By the end of my presentation the small congregation of Bethel was all on board and they gave generously to the efforts of Living Water International. When I came out to get Ruby and Peter Wheatland, Peter had tried to make a run for it. Fortunately, I had finally found someone slower than me and he only made it ten yards in the past two hours. I picked him up and we walked over to the parsonage for the night.

> *"I like running because it's a challenge. If you run hard, there's the pain - and you've got to work your way through the pain. You know, lately it seems all you hear is 'Don't overdo it' and 'Don't push yourself.' Well, I think that's a lot of bull. If you push the human body, it will respond."* **Bob Clarke**

Day 89 (Camdenton) 37 miles

"I stayed at Pastor Aliens' home, which was right on Hwy 54 so I didn't have to break down Ruby. I was thrilled to see Peter Wheatland had made it through the night, but I had growing concerns when Kate asked me if he was pooping. She said, "Rabbits poop a lot." He hadn't pooped a single raisin. The local newspaper editor came over before I left and we talked about how I was running across the country. I told him a few good stories and also told him about Living Water International and all the work that they were doing around the world. He was a nice, relaxed man and easy to talk with. He told me to stop by the office up the road on my way out of town so he could get a picture of Peter, Ruby and I in front of a well-known town building."

The weather said it was supposed to be thunderstorms all day. It had been raining in the morning but it subsided enough to convince me to squeeze in my 37 miles. Part of me knew running in the rain would be inevitable but I stuck with the happy positive side that told me, "Oh maybe it will blow over and it will be a bright sunny day." Either way, I was on the road again.

With about 16 miles to go, the storms finally caught up with me. It broke itself in slowly with a light rain and then quickly turned steady. I put the rainfly over Ruby and Peter's box. Peter looked like he had it pretty good all curled up warm, and dry. All the sudden it started to pour. I'm not even sure if we were still dealing with individual raindrops; it may have been buckets. It was one of those rains that windshield wipers can't keep up with. The heavens appeared to open the floodgates. With water pouring in my face, I struggled to even keep my eyes open. It was hard to even tell which way was up and which way was down because water was coming from all directions. It would fly straight up off of Ruby's spinning tires and fly straight sideways from the semi trucks passing at 60 mph. The shoulder of the road turned into a two-inch deep river. I'm pretty sure rubber soled shoes can't absorb water, but I was positive my rubber soles were saturated through and through. Practically submerged in water, I was drowning in the rain; however, it was not the rain that worried me but the lighting bolts that seemed to be flying past my eyebrows. For some strange reason I felt if I ran hunched over I could avoid one of these flying fire bolts of death from hitting me.

A lady pulled her red compact car over and asked me if I needed a ride. Standing outside her window, still dressed in only my t-shirt and running shorts and looking like I had just crawled out of a lake, I smiled and asked how she was doing. She seemed taken back for a second and then popped the trunk and said, "Here I'll give you a ride to the next town."

I said, "Thank you but if you can't give me a ride back to the very spot when the rain stops I have to keep running." As we parted, she was shaking her head seemingly unable to comprehend my motives, and I, soaking wet, felt too tired to explain them.

Immediately after she left, a huge lighting bolt hit the grey sky in front of me, and I regretted my decision. However, a mile down the road a small bus had pulled over and was waiting on the shoulder of the road. My vision was blurred with the massive amounts of water, but I assumed they were dropping off a child. As I approached, I saw no driveway. The back emergency exit door swung open to reveal two older ladies with big smiles. The one on the left shouted out, "Are you a runner?"

I smiled and said, "Yes." My second chance had come!

Just then, without warning, she slammed the door shut yelling, "Okay just checking," and they drove off.

I found the interaction strange. I thought about what could have been going through her head to just assume because I was a runner, I liked to be on the highway with lighting bolts flying all around and rain coming down in sheets. I figured she must have come to the conclusion long ago that runners like to do that kind of thing and that half of us are crazy. She must have been the kind of person who accepts every one for who they are. So in a way, it felt good to be respected, but I was still left standing in the rain with eight miles to go. A total of five people stopped their cars and asked if I needed a ride before the day was out. It was kind of everyone, but I was having a hard enough time convincing myself to run through the storm without people yelling out their car windows, "It's pouring rain and lighting, I'll give you a ride!"

The next day, clothed in plastic rain gear, Peter Wheatland, Ruby and I headed out to take on the heavy traffic and rain. The river along the road once again soaked my Asics running shoes making their advertised sixteen-ounce weight feel like 16 pounds. Only a few miles into the run, I came along a surprising building. The sign read, "Animal Hospital." Peter Wheatland was saved! Completely drenched and looking like a sailor in running shorts, I burst through the doors. In both hands I held the wet, slightly dead rabbit. The veterinarian took a look. He poked, listened and even squeezed a full Dixie cup of pee out of Peter. He said he had never witnessed a rabbit pee that much in his whole life. The conclusion, Peter had a broken back and was paralyzed from the middle down.

He only had partial feeling in his bladder. Peter's body was bruised all over and had a few broken ribs. The vet said the chances of him living were very slim. He gave me some tips on what he might like to eat and also how I could make him the most comfortable in his remaining days.

The road narrowed, leaving no shoulder at all. The afternoon traffic had picked up and the steady stream of cars made the road impossible to run on unless I wanted to play chicken with every single car. I figured I was relying too heavily on my rights as a pedestrian and stopped at the first motel I came to, frustrated that I had only gone six miles and that Peter was going to die any day. The rain seemed to be the appropriate weather.

Once again, I stood in my hotel room looking out at the rain and contemplating how bad the run would be. The hotel parking lot had a slope to it, and the low spots were made clear by the build up of water. Peter had shown big signs of improvement after his emergency doctor's appointment. I caught him eating and the evidence of him pooping was clear. I continued to gently squeeze his bladder like the doctor did until dehydrated pee would come gushing out. However, I couldn't tell if he was drinking his water or simply knocking it out of the container.

Eventually we were on the road again. Peter was tucked away under Ruby's rain fly. Both seemed to be oblivious to my navigational worries. Like I said, there was absolutely no shoulder and unfortunately traffic had not died down. There was, however, a middle turning lane that proved to be my safest option. As I ran down the dead center of tourist traffic, I sensed everyone looking at me. About eight miles later I turned onto Hwy 42 east. The traffic died down a bit but my middle lane disappeared.

This road proved to be equally frustrating. There was still no shoulder and endless curvy turns. In general, this wouldn't be an unusual circumstance, but the constant flow of cars and trucks that came around every corner forced me into the ditch. This pattern made it impossible to get into any type of running rhythm and the 26 miles took up the remainder of my day. I arrived in Iberia just before dark. The heavy traffic made sense after I realized it was graduation weekend. It appeared every single person in the surrounding area had come to the small school gym to celebrate.

Peter looked awful. I think the 26 mile bumpy ride had been rough on his bruised little body. He no longer lay with his head up but rather completely on his side making soft squeaking noises. The journey had become too much for the little guy. There is no doubt in my mind that Peter Wheatland was the only wild rabbit to ride 74 miles in a baby jogger across the state of Missouri. It was an accomplishment he could be proud of. I wasn't sure what to do with him; however, when I came across a beautiful field of wild flowers I decided it was a worthy place to die. I lay him down among the field's flowers. His big brown eyes half closed now stared up at the evening sky, my sweating face hovering over him. I wondered what he was thinking in those parting moments. Would he have thanked me or cursed me?

Leaving him behind seemed to be in both of our best interests. I would be attempting to find housing with someone at the graduation. The hair on my chin had already had grown long enough for people to question my good nature, and if I was also carrying a 95% dead rabbit along with me, it would have confirmed their doubts.

The valedictorian gave a classic speech about her family, friends and the road ahead. The death of Peter put me in a reflective mood as I listened. I felt that in that moment I could have given an inspiring speech about the road ahead. You can choose whatever road you want to travel down, but that doesn't mean it won't twist and turn without warning along the way. I would talk about how with the proper persistence and time you can be the person you dream of and go places no one has ever gone. How love will pull you through the dark time and how passion will guide you through the confusing ones. I would tell them all to find a dream that's bigger then what they currently think possible and to find the courage to take the first step towards it. No matter what anyone says, don't loosen your grip on that dream. Some will call it greatness, some will call it foolish, but all that matters is what you call it.

Everyone seemed pretty busy doing his or her thing and I wasn't in a talking mood having just laid Peter to rest. So I left the lights and people and walked down to the baseball field. I settled down in the visitors' dugout, probably because it felt less like trespassing. I knew it would storm tonight so I was thankful for a good place to rest.

The lights of the school made the wet field sparkle, and I sat on the bench thinking about Peter and drinking from a cooler of leftover Gatorade. The graduation let out and people filed from the building. Most of them walking by a happy student in a green or white gown. They would move on whether they liked it or not. A hundred young minds about to take on the world or let the world take over them. One after another they piled into their cars. The headlights would turn on and bring unknown objects into the light. As they turned the car toward the road the headlights would briefly cross over me but never stay. I was nothing more than a far off shape, an object, an unknown piece to the puzzled landscape that the graduates knew all too well. The distant chatter slowly died and the town fell silent.

CHAPTER 11
VISITORS FROM THE NORTH

I slept fairly well considering I was outside on a bench during a thunderstorm. When I awoke, it was still lightly raining, the kind of rain that you have to stare at a puddle in order to tell if anything is coming down. The nice benefit about sleeping outside was that I always started my run at first light. I had a long way to go today so being on the road at seven was a good move. I thought about going back a half mile to see if Peter had made it through the night. However, in the back of my mind I knew he hadn't, and I didn't want the image of his dead corpse in my head. Traffic had died down considerably compared to yesterday afternoon. I passed Hwy 133, which led straight up to Jefferson City. The light rain turned to a wet mist, that clings to you like dew and blends flawlessly with your sweat. Around every corner the rising and falling hills seemed to grow longer and steeper.

I jogged into Vienna after 12:30 p.m. The rain was starting to pick up and the clouds looked as if they were ready to burst open again. Since I had already come 22 miles I decided to see if I could find a library to hide out in and check the weather. After asking around a bit I located the library. Upon my discovery of the building also came the harsh reality of it being closed for the day. It was Sunday afternoon. Once again another local high school graduation was taking place and preparing to send their product into the world. I approached the scene dreading the thought of sitting through another random graduation ceremony for the selfish reason of being out of the rain. So instead I turned around and went back to the library, pulled out my sleeping mat and lay down. The four-foot over hang allowed just enough room for me to keep out of the rain. I slept for at least an hour before I put on my raincoat and walked over to a cafe to order some chicken.

The rain ended around three in the afternoon, and I was back on the road in hopes of making the 18 miles to Belle before dark. After Iberia, the traffic died down to almost nothing, and I was left running alone through one of the most beautiful valleys I have come across yet. It was so green; the tall wild trees had vines wrapping and hanging off them giving it a rain forest feel. A good-sized creek ran through the valley and crossed under the road several times. In most cases the trees were back at least a hundred yards from the road leaving room for wildflowers and tall green grass. A dozen horses bulleted across an open hilltop that elevated high enough to overlook the beautiful valley. With the storm clouds still in the backdrop, they ran at full speed down the hill and across another small creek sending the freshly fallen rain flying once again.

An elderly lady, who was clearly out for a Sunday drive, pulled up next to me and commented on the horses. She proceeded to drive next to me for the next three or four miles talking about the huge hills that were around the corner and everything else under the sun. She was naturally a kind-hearted older woman and ended up giving me five dollars for a burger and also calling a pastor in town to ask if he had a place for me to stay. Luckily, he did.

When I took off my shoes, I found the balls of my feet

swollen. This was most likely due to running 40 miles in wet shoes. Regardless of how it happened, it felt like someone had inflated my feet and I was walking on balloons. I attempted to ice them in the tub and stayed off them as long as I could. At noon, I walked down Main Street in search of a wireless signal to post my daily blog. I found one and was standing outside the building for at least a half hour. A policeman pulled up and asked if I was the guy that was running across the country. He proceeded to explain how he had pulled someone over yesterday who was out looking for me. The story was left at that since he couldn't remember who they were and I had no idea who it could have been. I would have to wait to see if I could find the other side of the story down the road.

A lady in a van also pulled over to chat. She had heard about the run through someone who had seen me. She told me she would report my story to the local newspaper. A man poked his head out of the building I was standing outside of and asked if he could help me. I think he knew perfectly well that by providing me with the Internet he was already helping me. I told him I was fine and he disappeared inside, which ironically, made my Wi-Fi signal disappear as well. I was done anyway and walked into the local newspaper office two buildings down. The reporter and I talked for a while before heading outside to take a picture for the story. The unusual scene of a tan long-haired man with short shorts pushing a baby jogger caused a lady, who was unable to take her eyes off the scene, to drive right through the stop sign adjacent to us. A truck racing down the cross road blasted his horn to bring the curious lady back to her guilty driving reality. The truck hit the brakes, the lady swerved, and the reporter and I looked at each other with wide eyes. The incident dissipated as fast as it materialized and life fell back to order.

I left Belle on tender feet and ran gingerly to Owensville a distance of only thirteen miles. The remainder of the day was spent icing my feet and keeping them elevated in hopes that the swelling would go down for tomorrow's 31 mile run. By morning my feet were practically back to normal. It's amazing how much better I feel after a day off or even a low mileage day of around ten to fifteen. The road into St. Louis had a wide shoulder that took a lot of stress out of the day's run.

I had been looking forward to St. Louis for a few reasons. Living Water had a network of over 200 runners who annually fundraised for them by running in the Go St. Louis Marathon. A meeting and tour of the Reliv head quarters was also scheduled. I had been looking forward to crossing the Mississippi River for some time now and officially ending my running days in the Wild West.

Day 95 (Wildwood) 34 miles

"The run has been twice as long, three times as wide open, and four times as challenging as I originally thought. It hurts to think back over the past few months; the barren desert, the freezing mountains, the endless plains and every single step in-between. I feel I have emerged from the West ready to take on the world, at least the world I am now familiar with. Every step that takes me closer to the Atlantic Ocean also took me closer to uncertainty. Even though running across America solo has been unpredictable and challenging, it's something I can't help but love. I have settled into and accepted this lifestyle. It's a great feeling when you know you're doing exactly what you were meant to do. During the endless miles I have played over future possibilities in my mind, but found them too overwhelming and have decided to focus on the task at hand: the east."

The road flattened out enough for me to make good time. Ruby and I knocked out a couple of the best mileage chunks we had run in days. I had been joking with people about Kansas and Illinois not giving Missouri any room, thus it got all bunched up and hilly. The thick woods appeared to be home for a variety of wildlife. Approaching a small creek that flowed under the road, I peered over the edge into the ditch; a large cement culvert created a small waterfall. Standing on the edge of the culvert looking straight at me was a bobcat. The moment passed in an instant, and we both went back to what we were doing.

I was looking forward to staying with Living Water enthuses and couldn't wait to get picked up and start meeting my very well organized line of people who would be guiding me through St. Louis. I waited at a Sonic for my host to pick me up. While I waited, I ordered their one-dollar dish. Ever since I had heard some ultra marathoner talk about how he would eat simple sugars like ice cream and soda at the end of his long races or runs, I have been hooked on a daily dose of ice cream whether it is the 49 cent cone from McDonalds or a gas station ice cream sandwich. I told myself I was just testing the theory but it may have grown into a habit. The combination of a good shoulder, sunny skies, not too many hills and a pick up from a contact made it a rare, uneventful and perfect day on the road.

Large cities could be daunting to navigate through and required a new type of skill set than wilderness survival. I have long dubbed this skill set "urban survival" and put this new set of developing skills to use whenever I passed through populous areas. The string of people that accompanied me through St. Louis made that hectic city a smooth sailing comfort.

Rachel and Rob were the start of the string of people who had run the St. Louis Half Marathon for Living Water. I would be staying with some of the participants throughout the area. Rachel was a personal trainer and her husband Rob was the youth pastor at the church, The Crossing, which has grown to over 6,000 members. They had both planned to run with me towards Wildwood but had to shuffle their children around so

Rob started out with me and Rachel ended up joining us after 13 miles. The furthest Rob had ever run was thirteen miles during a half marathon so when he pushed through the mental barrier to see how far he could go, it made for an exciting run. I think all runners who get into running long distances experience a run where they reach the tipping point. They finally push past their comfort level and familiarity of a certain distance and come out on the other side having gone further than they ever thought possible.

For me, I think it happened slowly and it is probably still happening the more I get into running the ultra-distance. I specifically remember a time when I broke my first mental barrier. I was in middle school at the time and school had been canceled due to a large snowstorm. During the storm, for some reason, I wanted to go for a run. With the snow falling, I ran from our house eleven miles to the top of what we called Suring Hill. It was a slow pace with a frigid windshield that pushed me past what I thought my limits were at the time. Since then, I have experienced many similar runs including the training runs that define a season or an off-season, the runs that help you forget, and the runs that make you feel alive. Watching Rob run the 22 miles and pass through the door into the unknown was simply remarkable.

On a side note, I just have to say their five-year-old son is my super hero. Ten days ago they got him a Spider Man costume with padded muscles and a mask. He had not taken it off since. Rachel said they were able to convince him to take it off at night, but other-wise he continues to wear it all day everyday.

Kerry Morgan, who I stayed with for a few nights, ran with me the last eight miles or so to the river. The goal of today was to make it to the St. Louis Arch that towered over the Mississippi River. In my mind it was the gateway to the east: a milestone. I always told myself during the training leading up to the run and over the last few months that if I could just make it to St. Louis, I could make it the rest of the way. I had come across some rough country that few people wouldn't even drive across, much less run across, and it felt good to have it behind me. However, 1,000 miles is still 1,000 miles, and I only had about six weeks left to complete the mission. I could see the arch from miles away. Our route led past the Cardinal's baseball stadium and under the arch to the edge of the great Mississippi River.

The following day Kerry and I meet up with two other men who had recently done the Go St. Louis half marathon with Living Water as well. We met downtown next to the arch and set out from there. They planned to stick around for two hours and rotate between running and driving the car. It felt overwhelming to cross the great Mississippi into the east. I was glad to have some running partners along, not only for the conversation, but also for guiding me through the city. As it turned out, they would also be my bodyguards through East St. Louis.

It was surprising to see how the development diminished the second we crossed to the Illinois side of St. Louis. Chicago spends up all the state funding and Missouri finds it silly to pour any money into a city that is not in their state. On one side of the city lay the St. Louis arch, parks, skyscrapers, the Cardinals stadium and the college. Across the border were run down gas stations, vacant lots and overgrown, torn up sidewalks. Less then a mile into Illinois, a policeman pulled over Kerry who was driving the car next to us. He was wondering if we were in our right minds running through East St. Louis. The policeman was sure to let us know that he had already wrestled someone to the ground just this morning. Of course, I had been completely oblivious to any type of danger. Everything looks simple on a map. The officer continued to try talking us into changing our route or riding in the car for a few miles until we got out of the neighborhood. He finally realized it wasn't an option and settled with escorting us through the city. The map called the road one thing but to everyone else it was prostitute strip. The neighborhood also had a record number of drive-by shootings, which had flared up again in recent weeks. We laughed about the situation as we set out running again, but as we reached the area our conversation stopped and our pace quickened. I have found what scares us most are the things we don't understand. Everything I saw I had already judged. I had preconceived notions of the people and the place and left it at that. I ran by it with blurred vision and left it how it was. The more I thought about how I didn't understand the people living there, the more I realized I didn't understand myself. Why would I act that way, why did I judge, why did I run?

"Any fool can criticize, condemn, and complain but it takes character and self control to be understanding and forgiving."
Dale Carnegie

There is a savvy sang that goes "the apple doesn't fall far from the tree." It was true that I had gotten a lot of my rambling ways from my father who was always bike camping, canoe camping, hopping box cars across the north to National parks or something crazy in his younger days. However, from this tree fell two apples. Although four years apart, they landed very close. My younger brother had brown hair, both lighter and straighter than mine. Although, also in top condition, his stature was bigger than mine. He was an inch shorter but broader and thicker in general. He was right-handed and I was left-handed and that may have been the only obvious distinctions between the two of us. His name is Joshua but he went by Josh for short and is what I would consider to be a best friend.

His life seemed to follow mine, only four years behind and with his own twist. We had virtually all the same interests and had spent our life to this point pursuing them together.

He was currently running cross-country on scholarship at the University of Wisconsin Green Bay, an undertaking that took me the greater part of four years to complete. We both have a dry specific sense of humor that, for an outsider is often hard to catch on to. We always seem to be on the same page usually only a few words and a certain look on the face could send us both to the hospital with a gut busting laughing disorder. Josh was always the first person I would tell my crazy ideas to because he simply would not judge them and instead helped me form the concept and think out the best way to go about it. Originally we were going to do the run across America together; however, his practical side told him to stay in school for the semester. Nevertheless, we always found time for a grand adventure even if it was only a day or two long.

We came up with the concept of run camping years before and tried it one weekend on the Door County Peninsula, otherwise known as Wisconsin's thumb. Josh had managed to convince his friend Kohner to come along. Josh is usually fairly relaxed about details in general and this proved accurate when explaining the trip. What Kohner thought to be a normal camping trip turned out to be something quite different. Leaving the car behind, we packed lightly, only carrying an emergency blanket, a few PowerBars, matches and some water.

Being in an oblivious state often heightens a simple adventure. These days technology and planning have eliminated unpredictability, which is what a true adventure is about in the first place. If you are looking for real adventure, then buy less gear and pack all your things in a bathroom kit. Spend less time planning; simply walk out the door and start. This is more or less what we did one late summer weekend. We ran 20 miles up the beautiful coastline, stopping only to swim, eat cherries from orchards and lay in the cool shade of tall oak trees. With the sun setting, we put the finishing touches on a drift wood shelter only feet away from the shores of Lake Michigan. That was when we learned that the "cool" factor and "practical" factor are two very separate categories when it comes to building a wilderness shelter. Lying on the rocks with nothing but emergency blankets and running shorts we fell victims to the elements. The wind would blow through the large cracks in the shelter and send all three of our foil blankets flying off. The loud noise of the useless foil blankets combined with the frigid temperatures and uncomfortable rock beds made sleep impossible. To date, that night still ranks as one of the worst nights of my life. However, it is also one of my fondest memories.

> *"The truest and most horrible claim made for modern transport is that it 'annihilates space'. It does. It annihilates one of the most glorious gifts we have been given. It is a vile inflation, which lowers the value of distance, so that a modern boy travels a hundred miles with less sense of liberation and pilgrimage and adventure than his grandfather got from traveling ten."* **C.S. Lewis**

Josh had decided to come down to St. Louis and visit me. He took his bicycle. I was not sure if he had ever ridden a bike more than 45 miles; however, the 550 mile trip from Green Bay, Wisconsin didn't seem to faze him. I never doubted he could make it physically, but I did have concerns about his bike. It was an old blue Austro Daimler from the seventies that I had left at college over a year ago. Josh found it in a snow bank last spring and greased it up. The tires were worn way past their lifespan. In some areas, the rubber was nonexistent and you could see the brown cloth beneath. The right pedal was slightly bent, making an unbearable squeaking sound at every rotation. Only the front brake worked properly, which he said, was fine because Illinois was flat as a pancake. He had camped the first two nights before a full day of riding in pouring rain convinced him to stay in a hotel. It turned out to be the kind of hotel where when you take a shower, the drain gets clogged with paint chips. I had not heard from him besides that he was coming my way and would meet me along the way. All the doubts I had of him actually making it at 9 p.m. on his fifth day were put to rest when I received a call from him asking if I could let him in the back door of the hotel. Leaving my room, I went down to the back door and immediately started to laugh when I saw him. I knew his state of exhaustion all too well. His crazy hair matched his selectively sunburnt body. The funniest thing of all was his deteriorating cargo shorts that he had worn the entire trip. It was good to see him.

"Brothers don't shake hands, brothers gotta hug!"
Chris Farley

CHAPTER 12
CHAFFING CURES

School was out for the summer so Kate drove down to visit arriving only hours after Josh's entrance. The old 1998 forest green Pontiac Grand Prix that we had bought from my friend for 20 bucks was not the ideal long distance vehicle but it was all we had, and it would have to work. I must admit I was sitting on the edge of my bed praying that her drive would go smoothly, half assuming it was almost certain she would be stranded somewhere along the way. The combination of darkness and my poor directions caused a frustrating last half hour of her driving around the hotel in circles. She had brought me a birthday gift even though my birthday was still a week away. If Kate is excited about something, it's impossible for her to keep it in. Still, standing in the parking lot my eyes were closed as I waited to be handed my birthday gift. In my arms she put something furry that had claws and sharp teeth... it was a new baby rabbit!

I convinced Josh to go on my morning run with me and we headed out of town. I was hoping to find a bike path that supposedly ran out of the town and connected to Hwy 140. Through a frustrating series of events and wrong turns we were right back where we started with a four mile loop under our belts. Our late start caused us to be running in the heat of the day. Later that day we found out it had reached a record breaking ninety-two degrees. We asked a couple different people for directions and finally made our way out of town. Eventually our different paces separated us by a couple hundred yards. Eight miles later the heat had taken the last of whatever energy we had to offer, and we called Kate to come rescue us. Our plan was to sit out the rest of the mid day heat and go back at it in the evening. This proved to be the best idea we had all day.

Once again it was a sizzling hot day. It reached 90 degrees in the afternoon. Thankfully we had learned our lesson yesterday, and I had Josh take me out to where we left off early in the morning to run a couple hours, then come back to claim my continental breakfast. He drove ahead eight miles and then biked back to meet me. It was the best I felt in a long time. With Josh coasting along on the bike, we flew through 12 miles in about an hour and a half. The plan worked out well and I was able to put together two efficient runs and sit out the heat of the day in comfort. We found a nice park by a small lake, lay around in the shade and went swimming. The new rabbit was glad to be out of the car and hopped around in the shade. Kate and I took the car into Greensville and got some chicken, corn on the cob and potatoes. However, cooking over an open fire on a ninety-one degree day didn't turn out to be the best idea, but it was a fun activity, and the meal turned out to be a success.

The days and miles passed by in their usual way. Josh and Kate had driven back to Wisconsin. I found myself miles down the road standing in front of a small town sign at 9 p.m. that read Welcome to Stilesville. The sun had long gone down, and my natural instincts of searching for a place to sleep for the night took over. Coming to a community park that had an elementary school right next to it I scanned the layout. The dugouts looked a little cheap and uninviting so I headed over to the playground. A small pavilion with picnic tables looked like a good backup if it started raining, but I was pretty sure it would be another clear night. Walking closer to the school I noticed it had a nice little court yard with flowers, well cut grass, and a few large trees. It was perfect. I pulled out my mat and sleeping bag and crawled in. With the combination of ideal sleeping weather, soft grass, and too many miles to count, I was out in an instant.

I slept well until a group of drunken people woke me up. They had stumbled onto the playground and were throwing little gravel rocks down all the plastic slides. As they ran around the playground, I hoped they wouldn't notice me lying in the shadows and give me any trouble. It was really the last thing I wanted to deal with. After 20 minutes my silent prayers were answered, and they disappeared into the night.

In the morning it dawned on me that it was Memorial Day weekend so I wasn't worried anyone would be at school, but I packed up and left at first light. I only ran seven miles before I stopped at a gas station that had a nice area to sit with big windows so I could keep an eye on Ruby. The gas station actually had wireless, so I stayed for a while before heading out to finish out the day. At the gas station I met a cool couple who had been a part of the Great Peace march back in the 80's. The Great Peace march was a group of 500 people who walked across America for Global Nuclear Disarmament. The journey took them nine months. The couple had actually met on the walk and married right after. It was a day of celebration as they reminisced on their 24 years of marriage and I on my 24 years of life.

A friend from high school, Andy, lived on the west side of Indianapolis and agreed to pick me up wherever I left off, so I headed back out on the roads for the push to Indy. The

Indianapolis 500 was in full swing, and the city was buzzing with excitement.

The next morning Andy dropped me off, and we parted ways. It was another hot summer day and traffic picked up as I approached the city. The sidewalks appeared and disappeared, depending on what type of stores or houses were next to the road. I was cruising along like usual when right in front of me dropped a power-line. The line had snapped off the pole and dropped to the sidewalk only yards from Ruby and I. As it bounced off the cement huge sparks shot out of the end like a firecracker. The line instantly coiled back to the pole as more sparks poured out from the top box. I had just passed a police officer parked a block back so I ran back to let him know. As he was pulling away I flagged him down. He had already known something had broken but didn't know the location. I told him where it was and left the scene with a crowd of people standing outside their businesses watching the sparks fly. Trudging down the road again I couldn't help but be thankful that I was moving at a slightly slower pace today.

Fox 59 news was trying to track me down and asked where I could meet them. We settled on a Long John Silvers that was a few miles ahead. Having already passed one Long John Silvers, I figured it must be the second one. I was 45 minutes early. A large poster of their new lemon slushy hung in the window; it lured me in and I ordered one and sat down to feed my must have now American culture desire. After running through the hot city it definitely hit the spot. It took me a full fifteen minutes before I realized I was still at the wrong Long John Silvers. The correct one was another three miles down the road. I had thirty minutes until the scheduled interview. As I approached the third Long John Silvers it was easy to see that this was the correct one. Fox 59 was waiting on the side of the road with the cameras and news van. Running up to the parking lot, I tucked the empty lemon slushy cup deep into Ruby's back pocket and smiled.

Molly Hartman was there waiting as well. She was an old friend from Fort Wilderness, the summer camp where Kate and I had met. I would be staying at her parent's house for the night in Carmel, Indiana, a few miles outside the city. I had fond memories of her brother teaching me to slalom on Spider Lake. It was good to be around old friends.

The next morning Molly and her friend Ashley were heading south to their lake house, so I hitched a ride, and they dropped me off at the third Long John Silvers. Ashley had just spent the last year on tour with Brittany Spears working as her makeup artist. She said before the tour Brittany's dad was looking for Christian people to put around her. The tour was also drug free, meaning they were constantly doing drug tests on everyone involved. She gave me a different perspective of the pop star than the tabloids, and I wondered how many other people we have judged without even knowing them.

The relaxing night came and went, and before I knew it, I was back out on the road dodging cars and dealing with my developing chaffing issue. There aren't many times I can remember being in more pain. The chaffing on my high inner thigh was simply torturing me. It burnt and itched like no other. I had chaffing in the past, but this felt like someone was sawing off my leg. I had no doubt it was because of running so many miles in completely soaked running shorts. Over the past week the record high temperatures had put my sweat glands into overdrive; it would only take a few miles before I would was drenched and dripping in sweat. I had plenty of skin protecting cream and bee's wax, but I had to stop every three miles or so to reapply. It was a distance runner's nightmare. The pain was so intense it would keep me up all night and I would often get up a few times in the night to search in the local bathroom for something that would relieve the pain. Anything.

I was going through the middle of Indianapolis, so the streetlights never ended. It seemed I had to stop every hundred yards. When I go through a big city, I have noticed that first there are the huge houses on the edge of town. Then you get into your average sized housing in nice suburbs. Then it turns into the rundown neighborhoods. When you pop into the city center it is usually pretty nice with a college campus, well-designed little parks and beautiful architecture. Then the process reverses itself on the way out.

Molly's mom had given me some aloe before I left, and I must say it saved my life. Every few miles I would stop, reapply and then cover the aloe with bee's wax. I don't think I would have made it if not for all the honks of encouragement from the city. Every other car seemed to recognize me from the news story that Fox did. A man drove by in a huge roofing truck with a trailer full of shingles; he blasted his big horn, leaned out the window and yelled, "Go Abe." Another man saw me running past his house and chased me down on his bike to talk to me and give me two bottled waters.

I was very grateful for the waters even though two boys playing in the street had done the same exact thing a few blocks ago. The excited boy said, "Well I didn't even know you were going to run down our road until we saw you." The truth was I was temporarily lost and was probably not even supposed to be on that road.

I was hiding behind a church sign applying my hourly dose of aloe and bee's wax when a couple of ladies spotted me and pulled their car on to the church lawn. They climbed out and started walking towards me. I quickly took my hands out of my shorts and put the cream away. They ended up giving me 60 dollars for a "nice meal." I'm not sure where they went out to eat but 60 dollars would buy me twelve foot long subs or 60 McChickens. Their random generosity was what kept me from being completely broke. The hotels had whittled my bank account down to nine dollars, so between the 60 from the two elderly ladies and a hundred-dollar bill that Kerry Morgan had given me, I had a bit of a cushion again.

Day 108 (Charlottesville) 20 miles
"The Ahearns have four boys and I know the middle
two. Eric, who is closest to my age, has been backpacking
in Peru for the last three weeks and will be out on the
trail for another three. Tom, the third son, is up at
Fort Wilderness spending the whole summer on staff
with my younger brother, Josh. The youngest, Scott
is sixteen and the only one at home now. We played
several games including 3d Tic-Tack-Toe, Carpet ball
and Ping-pong. He came out ahead on everything but
the Ping-pong. The games are a good distraction. Even
though the end is now with in reach the length of this
trip has worn me thin, I'm ready to settle down and
lead a normal life."

They dropped me off the next morning around eleven
a.m., and as I was getting the stroller ready, they asked me
if I would like to stay at their house another night. I believe
there is a certain amount of time you are supposed to wait
before answering a question like that. Perhaps a small pause
that lets them know you are thinking about it. I didn't allow
anytime for this socially acceptable pause to take place before
answering. "Yes." I loaded Ruby back into the car and took the
light version with me which included my fuel belt, a PowerBar,
water bottle, $20 bill and my cell phone.

As soon as she drove off I realized one more thing I would
desperately need, some kind of skin protection or aloe for my
chaffing. It is true that sometimes the smallest things bring you
down. My legs felt good; I had enough energy for a decent
run; but my goodness, life was miserable. I altered my running
stride to resemble that of a cowboy who had just gotten off his
horse. I did my best to keep from sweating, which is basically
an impossible task during a mid summer's day. About seven
miles into my run, I came across a Walgreen's and bought a
two-dollar jar of Petroleum Jelly. I felt a bit strange buying just

the Petroleum Jelly so I bought a big blue PowerAde and also got suckered into their two-for-one deal on Butterfinger candy bars.

I sat out on the bench and ate my Butterfingers, opened the can of Vaseline and took out a big glob. Right about then, a man pulled in and parked right in front of me. I patiently waited opening my second Butterfinger with my right hand and teeth while balancing the Vaseline with my left hand. I got a slightly strange look from the man but believe I covered up the situation pretty well. I was about to go for it again when a teenager walked out of the Walgreen's to his car parked slightly to my left. Again I waited. I glanced down for a moment to check that the chaffing had not ignited my running shorts causing them to burst into flames. As I waited for the boy to leave, I noticed a man from across the parking lot leaning on his rusted turquoise Toyota. He had a beard five times the length of mine and was smoking a cigarette, worst of all he was staring right at me. That was the last straw. I couldn't wait any longer. I threw everything in my plastic Walgreen's bag and jogged away from the busy intersection carrying my bag in my right hand and balanced the sliding oversized glob of Petroleum Jelly on my left. This distraction helped a little but I don't know if it was worth having the sticky Petroleum Jelly all over my hand for the next 13 miles. It was miserable to say the least. I don't know how I made it the 20 miles to Charlottesville, but eventually I was there, inflamed shorts and all.

Once again the Ahearns dropped me off, this time twenty miles down the road at the Charlottesville fire station, yesterday's stopping point. I headed down Hwy 40 once again. It was the National Highway's yard sale weekend so people had dragged all of their unwanted items out to the side if the road in hopes of turning it in to cash. The variety of things for sale kept the run interesting.

Four miles in, a man ran up from the opposite side of the highway. I instantly knew he was an ultra marathoner just by the type of hand held water bottles he was carrying. Runners usually don't carry water with them if they are going for a two-mile jog around the block. He ran up to me, introduced himself as Pat, and asked if he could join me. He said that anyone running down a highway with a baby jogger full of gear must be going a long way. As our conversation progressed I found out that he was leaving for the Western State's Hundred Mile in only a few weeks, one of the super bowls of ultra running. He had completed several other 100-mile races and just last month won a 24-hour track race in which he ran 116 miles. Stories from the trail made the miles fly by. I enjoyed hearing about his experiences of running the Leadville 100, Rocky Raccoon 100, Heartland 100 and other ultras that I had only read about. There is something very exciting about finding another person who has an equal amount of stories of one's body completely reaching its running limit. It was the perfect distraction to my chaffing pain.

Not too long after I met Pat, a few people from the NCAA Horizon League pulled up to get some footage of the run and a little interview. We ran on for about 20 miles before I said I was good for the day. He had agreed to let me crash at his house for the night, so his wife and four children came out to pick us up. We ended the run just in time because it started down pouring and lightning on the way back. I was very thankful my plans for camping had abruptly made a change for the better.

Pat did computer simulation work for NASA out of his home and decided to take the day off to run with me again. By the time we started running, it was incredibly hot, and the yard sales were in full force. Since we didn't have a driver, Pat ran about five miles with me and then turned back to get the car. He then drove ahead and parked the car at the Ohio state line and ran back towards me. When he caught back up to me again, I had reached Richmond. Once again, it was an entertaining run with all of the random roadside junk to look at and the abundance of interesting people these types of events seem to attract. A few people looked like they had come a long way to sell their stuff. They had 15 passenger vans packed with an assortment of things one might call "stuff". A big blue PowerAde from the Dollar General may not count as a treasure on the national highway to others, but it sure was to me.

My chaffing seemed to be under control, which was a relief, and I made good time over the 20 miles. I ran better when Pat was running with me. It's always is nice to have some help getting through bigger cities. In less than four hours, we reached the Ohio border. It had taken me just seven days to run clear across the entire state of Indiana despite some slow going around Indianapolis. I stayed a second night with Pat's family.

Pat's daughters were just starting to get into a few of my favorite instruments: the mandolin, violin and baritone ukulele so we had a good time trying to figure out a few tunes. I helped tune them up and gave them a few tips to get them going. As always it's nice to stay in the same house two nights in a row and get to know the people a little better. Plus Pat had a lot of Ultra running magazines to read that proved to be great inspiration for my remaining days on the road.

Day 111 (Dayton) 42 Miles
"We made an effort to beat the heat this morning and left the house a little after eight a.m. I'm definitely not known for leaving anywhere in a hurry, it's just too easy to put the blame on Ruby. One of Pat's running friends came over to tag along. I guess you know you're a runner when you can't remember a person's name but remember he ran a 4:41 mile and 10:15 two mile in eighth grade. Anyway, it was fun to talk about running and Pat's fat meat chickens on the way to the state line. We parked the car in a hotel parking lot and crossed into Ohio. We held a solid conversation and pace,

which usually happens when running in a group. They tagged along for over ten miles and then turned back for the car. I had no plans for housing so I trudged on most of the day."

Mid afternoon brought a heavy downpour that turned my running shoes into ten-pound dumbbell slippers. It was one of those summer rains you run in just for the experience. I remained shirtless and was not cold. As expected, a few people pulled over to offer me rides. After the bulk of the storm passed, a car stopped to offer me a ride, and as usual, I kindly declined. They drove off only to return a few miles later to give me the groceries they had just purchased. The mother said every time her children see someone in need they make her go buy them groceries. The children were also the first on my journey to ask for my autograph. I wrote it on an old envelope and the ink instantly smeared all over the place due to my dripping wet hair. As the car drove off I could almost hear their state of confusion, arguing if I was famous or homeless. The truth was that I was neither, but for the time being perhaps a little of both.

The sun came out, and I changed my socks and running shoes for my run into Dayton. Dayton, Ohio did not seem like a high-class place- at least the part I ran through. Garbage lay along the streets where abandoned buildings stood. Run down front porches were filled with empty liquor bottles. A man pulled his van up next to me and rolled down his window, lowering his shades he said, "You have beautiful legs."

Not quite knowing how to respond, I half smiled and said "Thanks," which was followed by a forced laugh. I had hoped to camp the rest of the way to save some money, but the rain began to fall again, and the comfort of daylight was departing. The long 40 plus mile day left me wandering lost on the streets of Dayton wishing that my stolen GPS would magically return. With darkness approaching, I jogged up to the bright safe lights of a gas station to check out the map. My shoestring budget never justified actually buying the map because I always figured that a new state was just down the road anyway. Standing motionless in the shadows of the gas station's white lights I stared down the dark road that led east,unwilling to go on, unwilling to stay.

Top - Peter and Ruby the only ones managing to stay dry.

Middle - Running shirtless in the heat of the day

Bottom - A simple motel room after my first 50 plus mile day.

CHAPTER 13
THE DISTANCE BETWEEN US

I had a good morning run of eight or nine miles to Eaton on a bike path that ran parallel to Hwy 35. A McDonalds right off the bike path lured me in with my discovery of free wireless internet. I sat next to the big windows so I could keep an eye on Ruby who sat a foot away from me on the other side of the windowpane. Outside the window, a sidewalk crossed the exit to the McDonalds and down the cement path rode three boys on cheap mountain bikes. I noticed the boy in front because of his plump round body covered by a bright red t-shirt. The event to follow was one of those situations you see unfold and it feels like it is in slow motion but you have no ability to stop it.

A black SUV flew through the parking lot towards the exit. The group of three boys coasted down the hill on the sidewalk crossing the bike path and headed straight for the exit. The plump boy dressed in red looked back at his two friends who followed close behind. The man in the SUV unwrapped his big

Mac and studied where to take his first mouth-watering bite. A moment later, their worlds collided and the boy bounced off the big vehicle like a rubber dodge ball. The moment came to a screeching halt, and the boy lay motionless on the sidewalk. In an instant a swarm of people gathered around the boy, half of them on cell phones and the other half covering their gaping mouths. A police car was the first to appear followed by a second unmarked police car and then finally an ambulance. I left the scene before it concluded and the story remained unresolved in my mind as I ran on.

I was confused when I reached the next town. It was supposed to be Jamestown but instead everything read Cedarville. What I failed to realize was the bike path split out in four or five different directions from the town center. I had been running down the wrong path for nine miles. Jamestown was seven miles straight south. I felt like an idiot. Between last night and now I had made no progress but instead have been zigzagging across the map, making huge circles and taking the most indirect route possible from point A to point B. Tired and depressed, I started walking south. I never feel like running after these incidences. The sun was starting to set, and I knew I would have to sleep in a field or something.

Just then a lady pulled up next to me and asked if I needed any help. I was blunt and asked if she lived around here. She pointed to the house a hundred yards back. I asked, "Can I crash at your house?"

She didn't even hesitate and said, " Yes." I can't imagine what it must take for someone to welcome me into their house with my appearance and them not knowing anything about me. There is always this awkward feeling I get and wonder if they think I'm some sort of crazy escaped convict. Knowing these thoughts are probably going through their mind always makes things interesting. She said she had a peace about asking me, so maybe God knew I needed a little pick me up.

Day 113 (Wayne Township) 33 miles

"I started running early this morning and ran south into Jamestown. Once back on my route I magically felt better and got in some good mileage on my way to Washington Court House. It felt like a long way to run. A lot of people have been talking to me about the end, which is exciting, but for some reason I have been freaking out. It doesn't seem like the end at all; in fact it feels exactly the same as the middle. A single mile is still a mile long and each day on the road is still twenty-four hours. There have been so many highs and lows on this trip. One minute I'm surrounded by people who are excited about the run and the cause and the next I'm alone again with only miles in front of me. I don't mean to be a downer about everything, but today was just a down day. It seems my emotions are fragile, and something like my phone not working drives me insane. I often pass the time by talking to people while I run. I usually talk to Kate at least three times a day and without that connection to familiarity I feel lost inside a goal that's too big to see out of."

When I reached Washington Court I was able to get online to update a few things and talk to Kate on the chat. I told her to come which was bad of me. It would be a ten-hour drive in the old green Pontiac and she was busy with work and school. I told her I was joking and left Washington Court on a bike trail headed east. As I left my broken phone rang. It was her. The frustration of my broken phone set in- she could hear me, but I couldn't hear her. I felt isolated, cut off from the world. I talked to her for at least an hour, not even knowing if we were still connected or if she had hung up. I asked her questions like always except now I tried to predict the answers. I told her the long boring stories that I normally wouldn't if I could hear her. I eventually said goodbye and goodnight not knowing if she was there or not.

The bike trail turned to gravel when the sun went down, and I went another half hour or so before looking for a place to sleep. I found a small paved road that crossed the trail and walked down it until I got to a house. A dog started barking from behind a paint chipped wooden fence so I walked to the next one. It was a farm; farmers usually are out and about pretty early so presumed they would find me. I walked to the next one. It was perfect; a creek ran through the side yard of the single story ranch house. Sleeping in mowed places is much more desirable than the woods. Setting up my tent hammock in the dark I didn't bother with the rain fly since the stars were out. That night I had a vivid dream of a little girl who lived in the house. I dreamt that she came outside into the yard and saw me sleeping. She instantly ran back into the house and told her mom. A moment later they invited me in.

I woke with the sun, as I usually do when I sleep outside. In my opinion it's a good natural way to wake up. I had gone to bed in nothing but my running shorts and a t-shirt, which was fine until the early morning hours when the night starts to forget that the sun ever existed at all. I had gotten out of my hammock and put on some warmer clothes and also put my sleeping mat down under my sleeping bag, which usually helps with insulation. It turned out to be a good move, and I slept comfortably for the rest of the night. After packing up the hammock, I headed back to the bike path to finish out the remaining miles of gravel trail. When it turned back to blacktop, it was easy going. About an hour into it, I lay down on a bench and took a little nap in the morning sun. I had been pushing hard attempting to make up mileage over the last couple weeks and the 42 miler may have set me back a bit in the energy department. I had been thinking of our wedding all morning and worrying about how post run across America life would work out. Time would only tell I guess, but it would be nice to figure out some kind of plan. I needed to find work to pay rent on the apartment Kate and I were hoping to get.

The trail ran past a small park in Frankfort, and I stopped to fill up my water bottles. I had only a few ounces of water left since the end of last night's run and was on the edge of dehydration. Pretending to be a basketball star, I shot an empty can of Reliv at a garbage can that sat ten feet away. It bounced off the rim and rolled across the parking lot. My head dropped; now it would take twice the effort to throw the thing away. This is pathetic, I thought to myself.

Walking along the baseball field fence back to the trail, I heard someone yell "Abe" from across the park. At least I thought they said "Abe." It must have been "Hey," though. I stopped in my tracks and tried to make out the figure that was swiftly running down the hill. It looked like Kate, but I didn't let my mind go there; I would just be disappointed. Besides, I had just talked to Kate last night on the computer and she was over ten hours away back in Milwaukee. Maybe mirages are a nice benefit of being dehydrated, and I decided to let the mirage go as long as I could. My heart stopped beating and then resumed at twice the tempo, still frozen in my tracks; there was no need to deny it any longer; it was her. I left Ruby and ran to her; we meet at center field and threw our arms around each other spinning around until we dropped to the soft green outfield grass. I held her face and looked straight into her eyes. On a sunny day two brown freckles appear in her left eye; it was her, she was real, and she was here.

She had canceled all of her obligations for the next couple days and driven all night to get to me. She had listened to me rabble on last night on my one-way phone, taking clues from my blabbering to narrow down my location. She spent the last four hours driving around looking for me and asking anyone who would listen if they had seen me. In all of the excitement of finding me, she locked the keys in the car, but we didn't care. We had both gone from the depths of discouragement to cloud nine. We left the car and walked to an old cemetery where we stayed for the rest of the day laying in the shade of a big pine tree. Needless to say I didn't run another step; there would be plenty of time for that tomorrow. Today life was perfect.

The next day Kate went ahead and hung out in Chillicothe and waited for me to complete my morning run. Between the short day yesterday and Kate waiting at my destination, I covered the ground in good time. I followed the bike path most of the way until it ended at the Hopewell Indian mounds. Apparently these large circular mounds of dirt were used as a gathering place for area Indian tribes. Chillicothe was a cute town with a classic park conveniently plotted at the end of Main Street. After my morning 25 miles, we hung out in the town's coffee shop for a while then walked down to explore Yactangee park. The park had a big pond surrounded by willow trees and filled with swarms of white ducks and swans. It was wonderful having Kate around during the downtime and made exploring the areas I ran through more enjoyable. We sat under a big willow tree as the sun set and watching the people pack up their picnics and toys and head home for the night. The swans made their final lap around the pond and then hopped up on to the soft green grass for the night.

Our time together flew by and eventually life ripped us apart again. Kate had to be back in Milwaukee for a meeting in the afternoon so we parted ways at 5:30 a.m. It felt good to be on the road before the sun came up. Mentally, I was well rested and ready to cruise through the next 600 miles and wrap this thing up. I knocked off seventeen miles before 9:30 and figured I would hang out at the local library until the summer sun cooled down and then head out again in the evening. My plan was to take a shortcut on the back roads, but as I crossed the highway and looked over my shoulder I saw a huge black cloud right on my heels. The cloud looked like it was ready to burst at any moment. The wind picked up and I ran to the closest building. It was a Family Dollar with a good sized overhang. The rain poured down and the thunder roared making the storm all the more intense. I sat next to Ruby contemplating my options. I didn't mind running in the rain if there was a set destination, but I was heading into miles and miles of back roads. If I left the town I would be camping and regretting my decision within a few miles.

My legs felt good, and I really wanted to try and make up the distance I hadn't traveled in order to hang out with Kate. A positive aspect about really bad weather is people feel sorry for you and you're always offered help. That's exactly what happened. A man came up to me, and we talked a little bit. I didn't ask for anything but he was committed to helping. He would ask me something and then disappear into the store to talk it over with his wife, then come back out and ask me another question. He said he lived four miles down the road and that I could come over and have dinner while I waited out the storm. He was a cement worker and had raised three boys on a small Ohio farm. He loved to trap and hunt in his spare time. His hunting habits must have worn off on his boys because he had over ten of the biggest bucks I have ever seen mounted in his home. Each one had a story behind it. His wife cooked some vegetable soup, and I had my fill of baked goods that were leftover from the rummage sale that no one came to. When the storm passed, it was dark and he offered me a bed and room for the night. I no longer had the desire to cover any more miles for the day and decided my efforts would have to wait until morning.

Day 117 (Parkersburg) 44 miles

"I wanted to get a good start on the day's run for two reasons, it was supposed to start storming at ten a.m. and continue on throughout the day and second of all, I had a long way to go. I got dropped off around 6 a.m. and turned onto the winding hilly back roads. The rain had left the air muggy and hot but it was far better than running in a rainstorm."

My morning goal was to make it to Coolville, a distance of 25 miles. The back roads were absolutely crazy and I can't believe I didn't get lost. It was a network of small paved roads that seemed to branch off and flow around every hill. The back roads were not on my map so I ran with the sun in my face hoping I would hit Hwy 50. Right when it started to rain I reached the Coolville library. It was around 11:30 a.m. I was excited I had executed the morning run perfectly after my failed evening run the day before. The library was closed so I sat outside under the overhang for three hours resting and waiting for the rain to stop.

At 3 p.m. I got my lucky break and headed out for an evening push to Parkersburg, West Virginia. Depending on where I stayed it would be around 15 miles. The first hour or so it lightly rained, and it felt good to run in the mist. As I approached the Ohio River the hills doubled in size and my pace took a turn for the worse. I crossed the Ohio River on the freeway bridge and felt I was stepping out of the Midwest and into the East. West Virginia is beautiful. I took the first Parkersburg exit, which ended up being a huge mistake because it took me to the top of a bluff that overlooked Parkersburg and the Ohio River Valley. In hindsight I'm glad I got to see the sunset view of the valley, but in the moment all I could think of was the two extra miles. I thought of camping in the park but had hopes of finding the Assembly of God Church in town. The possibility of sleeping indoors to avoid the approaching storm would be the ideal situation.

Parkersburg seemed to have a million churches and nobody really knew which one was which. Darkness fell on the town, and I found myself wandering around on the streets. I came to a locked church that had an open wireless network and spent the next two hours trying to download Skype in order to make some phone calls. I never got a hold of the church but managed to set up a meeting spot the next morning with the news station. It was around 10 p.m. and a car full of young high school boys sped into the parking lot in a dark blue Honda. Two of them jumped out and popped a few of the Vacation Bible School balloons that decorated the church entrance. They saw me and jumped back into the safety of the car staring me down as they drove away. They must have decided I wasn't a threat because they drove past a half hour later and threw a large Wendy's soda at me. The big cup of liquid hit the brick wall I was leaning against and showered down over me. They waited for my reaction, but I gave them none.

I crossed the parking lot to a small grassy spot. The teenagers drove by again and to my regret, saw where I had moved. I probably should have just left the area but was too tired to go another step and just wanted to sleep. It had been a long day between the high mileage, rain and hills. I was about to drift off when my rain fly got hit by a couple of wet hotdog buns. I barely noticed but it was enough to keep me from the sleep I yearned for. The harassment continued and escalated to rocks and laser pointers. It dawned on me that I would have to confront them, it was their summer vacation and they seemed to be begging for some excitement.

I peeked out from under the rain fly to see the oldest boy pointing and motioning a plan to sneak around the backside of the church to get me from behind. I slid into my black pants and running shoes and rolled out from my sleeping bag shirtless. They had set their plan into motion and were gone. I snuck across the parking lot to the corner where I thought they would be coming from and positioned myself. I stood in the shadows of the street lamp shirtless, put out my best cold-blooded killer vibe or X-men Wolverine look.

The three boys came around the corner and there I was, staring straight at them with all the rage and intensity I had. I have never seen anyone look so scared in my life. They completely lost it and took off running screaming like little girls. I went after them. We flew around the corner of the church, and I easily gained on them. The weakest one screamed the loudest because he was falling behind. His untied high tops and baggy jeans were slowing him down along with his pathetic running form that included a swiveling head, which kept looking back at me. I easily gained to within five yards as he looked back, I growled like a rabid wolf. His legs couldn't keep up with the rest of his body and his shirtless little frame tripped. At full speed he bounced off the pavement, crashed into a guardrail and then rolled down a steep hill. The other two boys had split as I hurdled over the fallen prey and chased the ringleader up the hill. He led me down a back alley, and I had settled into a pace that would last for miles. He noticed my steady, relentless gain and stopped in front of an open back door. The boy turned and screamed something at me in pure terror that no one could have understood. When he realized I wasn't giving up the chase, he disappeared into the house. I stepped into the door after him. It was probably one in the morning at this point, and I figured it was the boy's house with his dad asleep on the couch. However, the man sat up still half drunk and with an expressionless face looked right at me and said, "Hey some kid just ran in my house." I said nothing and left.

I wish I could say the rest of the night went by without any more troubles but it didn't. The weather was just getting started. The mosquitoes were out tonight so I zipped up my sleeping bag. A couple minutes later I got so hot I stripped off the rest of my clothes and slept naked. I was sweating and not sleeping well at all. As if the night couldn't have gotten any worse, it started to pour. Rain flew in from the sides of the rainfly as the water built up in the corner. After running forty-four miles, I lay naked and awake in the dark pouring rain. It didn't matter at this point in the journey. Nothing could phase me; I felt nothing.

Above - After a run in Illinois, temperatures exceeded 95 degrees.

CHAPTER 14
THE NO GOOD SAMARITAN

The rain stopped, and I could hear birds starting to chirp. The long night was over. I picked my head up to look out from under my rainfly and then quickly set it back down. I noticed a small screaming sound whenever the motion was repeated. I did this several times to try and figure out what in the world it was. I finally discovered the largest beetle I have ever seen burrowed in the long sleeved shirt I was using for a pillow. It was a Lucanidae Beetle, which can grow up to 37 mm long. My spine squeamishly shivered thinking about my head laying on it all night. I shook it out and decided anything, even running, would be better than lying in the grass for another hour. I packed up and headed to a McDonalds three miles away on the outskirts of town to wait for the news crew. At 9 a.m. they did the interview and a moment later I was off and running again.

I had made sure to check the weather this time and saw that showers were likely to take place in the afternoon, so I wanted to get in as many miles as I could in the morning. Unfortunately, Ruby had another idea and picked up a piece of glass in her right tire. The tube instantly went flat. A 20-inch bike tube would either be 80 miles ahead in Clarksburg or five miles back in Parkersburg. I couldn't stand the thought of having to backtrack five miles so I pushed on. The hills were getting larger and larger the further east I travel. They rise up for a mile or more and then back down. I took the hills one at a time; progress was slow at best. I got to Ellenboro just before sunset and went to the local McDonalds. It seemed to be the hotspot in the small town. Craving salt, I ordered a large fry. I asked where I could find a hotel and the man said there might be one three miles down the road in the next town.

I decided to go for it. I ran out of town along old railroad tracks that were parallel to the highway then jumped on a side road. Before I knew it, I had lost the highway and was wandering around in the dark on the back roads of West Virginia. Three miles went by and there was nothing. The hills were like the relentless waves of the ocean, rising and falling in rhythm. In these hills you could literally be a half-mile away from a big town and not even know of its existence. A lady was out on her porch and so I called down the steep hill to her, "Do you know where the hotel is around here?" She didn't say a word but instead walked inside.

A shirtless man came out and asked what I wanted. I repeated the question and he yelled back, "You ain't gonna find nothin' around here honey. Better watch out the dawgs are comin' after you." Sure enough two barking dogs were making their way up the step incline after me.

"O okay, thanks," I quickly said and scrabbled back up the steep hill grabbing chunks of grass to keep my balance. I got to Ruby before the dogs reached me and disappeared into the night before they reached the top of the hill.

I eventually found the highway again. The traffic lights looked uninviting, so I decided to just suck it up and sleep outside again. There was a nice mowed spot the looked like public land. It was next to some kind of water treatment facility with a chain link fence and barbed wire around it. The fence was perfect for setting up my ground hammock. I had learned my lesson from the night before and set up the whole hammock instead of just the rain fly. It was still hot, so I placed my sleeping bag under my mat for some extra cushion; it was a good set up. The 37 mile day felt a lot longer due to my lack of sleep the night before.

Right when sound sleep was finally mine, a car pulled up on the other side of the small building. The bright headlights woke me, but I was too tired to even look out to see what was going on. A deep voice called out, "Step out of the tent sir, this is the state patrol."

I rolled my eyes, of course it was. "Just getting some sleep sir I'll be gone in the morning."

As if I had said nothing at all he repeated his statement "Step out of the tent sir, this is the State Patrol." I didn't move a muscle and lazily thought to myself, maybe he would take me to jail and I could sleep there for the night. I would have a nice bed, four walls and a roof. The thought was tempting but instead I sided against my instincts and scrambled out of the small cocoon. His Maglite blinded my eyes as I covered my face with my hand. There was no need to search me because I wasn't wearing much of anything, but at least it was more than the night before. He said I had set off some kind of alarm by messing with the fence and said I would have to move on.

He hovered over me with the flashlight as he watched me take down the set up. With the mat and sleeping bag still in the hammock I bundled it all up in my arms and stood next to Ruby looking at him. I must have appeared to be completely naked when holding the bunch over my running shorts. "Is that good?" I asked. A bit set back by the situation he searched for the correct response.

"That should work," he replied.

"Well good, I'm glad we got that all sorted out." I turned around to the wooden fence directly behind me, plopped down the entire mess and began to set it up again. I wasn't sure why his light was still on me so I turned around and looked at him again, waiting for an explanation.

"You're going to set up right here?" he asked.

"Yep, not going any farther than I have to," I said as I commenced to setting up the tangled equipment.

"Well that should be fine I guess, as long as you're not touching the fence." His once firm and direct words were now filled with ambiguity.

I assured him I would stay clear of the fence and with that I was left alone again. His bright headlights disappeared over the hill, and I stood in the dark staring at the tangled mess. "You have got to be kidding me." I muttered to myself.

When I finally got to sleep, I slept very well. I didn't wake up until 7 a.m. which is very unusual for me when I sleep outside. The hills of West Virginia were breathtaking in the early morning fog. The morning rush seemed to be on across the four lane Hwy 50. Hwy 50 is my route across West Virginia, one of the few roads that run straight through. It had a nice shoulder, but the hectic traffic would remain until Clarksburg. I had high hopes of making it to a hotel in Clarksburg, a distance of 36 miles, but my energy level and Ruby's flat tire were not on the same page. Yesterday the flat tube had developed a huge knot causing the tire to bump at every rotation. It was driving me a little insane but I assumed it was just the unavoidable air pin. When I realized the air pin was attached to the tube itself and the whole tube was useless, I took it out and proceeded with only the tire and rim. This helped but Ruby was still lopsided causing her to constantly veer to the right.

A large antique store sign caught my eye but it was the small letters reading "ice cream" that held my interest. As I crossed the road three boys pedaled towards me and stopped to talk. They looked in equally rough shape. They had left from San Francisco 23 days ago and were riding their bikes to their friend's house in New York. We seemed to relate well to one another. I could have kicked myself as I sat on the guardrail and watched them pedal out of site realizing they probably had a tire repair patch I could have used. My own stupidity would cost me at least 20 more miles of hard travel. There was only one thing that could lift my spirits: ice cream.

I was back on track and had just climbed a mile long hill when a man pulled over from across the freeway, and yelled that I had forgotten my wallet at the antique store. I tried not to think about the added mileage and ditched Ruby in the woods to run back down the long hill to the antique store. Sure enough I had left my life lying right on the table. There was no way I was going to run back up that hill again. I may be getting lazy, maybe smart or maybe just less shy, but I stuck out my thumb to hitch a ride. An old truck pulled over and I told him I just needed a ride to the top of the hill. I jumped in the bed and rode up the hill.

A motel sign that pointed towards Salem convinced me I had gone far enough for the day, and I walked off my route into Salem. As luck would have it, the establishment had been shut down for years. It was around sunset at this point so I turned to a church to rescue me. I saw the pastor working on his computer through the window, but by the time I got my nerve up to ask for help he had disappeared. I knocked as loud as I could on the door but no one answered. Across the parking lot was the parsonage and I tried that. His daughter who was around my age opened the door. I explained my situation, which I have recited a thousand times since the Pacific Ocean. She said she would try calling her dad. I sat in the parking lot and waited. Fifteen minutes went by and finally the girl came out to a car that was parked outside the house. Being careful not to look at me, she grabbed her cell phone and disappeared back into the house. Five minutes later the pastor finally stepped out of the big white church and onto the porch. I went through the process of introducing myself again and asked him if I could crash in the church or if he had any other ideas. To my disbelief, he said, "Well we don't let anyone stay at the church." I had run over a hundred miles in three days and had not had good night's sleep in at least two nights. His response put me over the edge and I glared in his eyes and said "Who's WE? You and God? I stared at him waiting for his own words to reach his brain but they never did.

Eventually after a long silence he couldn't handle my awkward stare any longer and said, "Sorry I don't know what to tell you." I asked again if he had any other housing ideas and he repeated himself, "Sorry I don't know what to tell you."

As I walked away in the dark, I wondered what I would have thought if I weren't a Christian. What if that had been my first encounter with a church in years and I had needed help at my lowest moment? How come a concrete worker can walk across a parking lot in the pouring rain and offer me a bed, food, a ride and 13 dollars for groceries, but a "man of God" standing in front of a huge white church with his house twenty yards away can look me in the eye and say, "Sorry I don't know what to tell you." All I knew is that I would rather work along side the concrete worker on a Sunday morning than sit in that Pastor's white church and listen to him blabber on about God.

I slept in the park across from his house. It rained all night. In the morning I placed a Living Water brochure in the door of his church. I wrote on it.

Dear Pastor, you should read this Luke 10: 30-37.

"A man was going down from Jerusalem to Jericho, when he was attacked by robbers. They stripped him of his clothes, beat him and went away, leaving him half dead. A priest happened to be going down the same road, and when he saw the man, he passed by on the other side. So too, a Levite, when he came to the place and saw him, passed by on the other side. But a Samaritan, as he traveled, came where the man was; and when he saw him, he took pity on him. He went to him and bandaged his wounds, pouring on oil and wine. Then he put the man on his

own donkey, brought him to an inn and took care of him. The next day he took out two denarii and gave them to the innkeeper. 'Look after him,' he said, 'and when I return, I will reimburse you for any extra expense you may have.' "Which of these three do you think was a neighbor to the man who fell into the hands of robbers?" The expert in the law replied, "The one who had mercy on him." Jesus told him, "Go and do likewise." **Jesus**

The past four months had turned me into a running machine. Hardened by miles and weather the steep mountains of the east were no match for the tan rippling quads that did what ever I asked of them. My dad came out to tag along for a few of my final days on the road. We enjoyed the Appalachia culture of large families sitting out on their front porch as well as camping in Lenny's Trailer park with Dad's tent (which had no poles) and meeting the numerous interesting people along the route. Also, the famous bike race Across America passed us on Hwy 50. The participants are fully supported bike racers that speed cross-country with little rest. The winner usually won the race in as little as nine or ten days.

Day 121 - 29 miles (Grafton) camped
Day 122- 29 miles (Maryland state line) motel
Day 123 - 30 miles (New Creek, WV) camped
Day 124 - 29 miles (Augusta) motel
Day 125 - 25 miles (Heyfield) motel

The Winchester high school cross-country team was out on their first training run of the summer. The scene looked all too familiar. Fifteen to twenty young skinny students were running around the grassy school grounds. I stopped by and was able to talk with the team and share a couple tales of the west. After the encounter with the team, I hung out in the library for a few hours trying to figure out the route ahead. Hwy 7 from what

I saw was the best way into Washington D.C. The four-lane highway had no shoulder and was quite a pain. The traffic got so bad I decided to wait until night when it died down. A few stoplights on the four-lane road caused cars to come in waves of thirty-to-forty. Ruby and I would be forced into the ditch when the wave drove past, and then sprint out of the ditch like Mel Gibson in Braveheart screaming down the temporarily empty highway as fast I could. My eyes were peeled for a place to sleep for the night but as the miles passed nothing stood out. When darkness came over the foggy hills, I dropped in the front yard of a huge mansion. With its numerous gardens, neatly trimmed hedges, water fountains, and carpet like grass you would have thought it a great spot to spend the night. However, there was a house dog that became aware of my presence a hour after I fell fast asleep. Between the endless barking and my spot being too close to the highway, sleep was sparse. I was too exhausted to care.

Day 127 - 44 miles to Vienna, VA

"This morning at 3 a.m. I decided I might as well be running if I wasn't sleeping. It turned out to be a great plan because hardly any cars were on the freeway. I passed a bored policeman and waved. It didn't take too long before a couple of his friends tracked me down. Two policemen pulled me over around 4:30 a.m. I was thankful I had sent my ten-inch bar knife home with Dad because once again they wanted to know if I had any weapons. They ran my ID and told me about a bike trail that ran along Hwy 7 into D.C. At 6 a.m. I found the trail and took a nap until 8 a.m. My early morning excursion covered sixteen miles and also scored a car free route that lead me straight into D.C. I can almost smell the ocean!"

"And in the end it is not the years in your life that count, it's the life in your years." **Abraham Lincoln**

CHAPTER 15
THE ATLANTIC OCEAN

It was a hot day, and the miles went by slowly. Most of the people on the trail were bikers. It was excellent people watching, and I enjoyed the change of pace. A passing biker is a lot less intimidating and deadly than a passing semi. When I reached Vienna, I asked around for directions to the Metro Train Station. My plan was to ride the train into D.C. and stay with Linzy, a friend from High School. I know Ruby and I have been through a lot together but man she makes me look like an idiot in public sometimes. It was an interesting ride, and of course I lost my metro ticket on the way and had to explain my story to security. I was very thankful Linzy came to my rescue. Even though I feel I'm an expert at urban camping, the big cities are still a challenge. It was a long thirty-six hours, and I was happy to have a place to stay for the night. It had been a long hard push to D.C. I covered 214 miles over the past seven days.

I didn't get going until 3 p.m. when I left the air-conditioned apartment I was astounded by how hot it was outside. I felt more confident riding the subway back to Vienna. Last night's issues had made me a subway vet and the trip went by without a glitch. I had left Ruby back at Linzy's and planned to pick her up after my run. I decided to take the bike path into town vs. the roads even though it would be three miles longer. I was more at home on the bike path than the busy highways and intersections. I ran surprisingly well despite the heat. Twelve miles into the run I got another bloody nose thus completing my unwanted goal of having bloodstains on every shirt I had. I was a bit concerned I was bleeding from my lungs when I started spitting out blood, but I'm pretty sure some just went down my nasal passage and into my throat. However, it was hard to be certain; it's probably just better to stop running for a bit when you get a bloody nose.

I was scheduled to meet Michael Plato, a Living Water staff member at 7:30 at the Lincoln Memorial and thought I would beat him there until I realized I had taken a roundabout path that added on a few miles. Michael ended up riding his bike out to find me. I was thankful to have a guide to the Memorial. After picking up Ruby, taking a quick shower and finishing a call in radio interview, we headed over to a barbeque where I met some incredible guys who are big supporters of Living Water. Most of them have been on mission trips or supported trips for the organization. I must say it was quite a sight to come over a small hill and see the Capitol building and the 555-foot Washington Memorial rising above the banks of the Potomac River. The memorial is the world's tallest stone structure and houses many of the national gifts that are given to America. Michael said that the reason for the change of color in the brick half way up was because they stopped construction on the monument to fight the Civil War. When they finally got back to building a different colored brick was used.

Day 132 - 27 miles
"I ran very well today despite it being hot. I got another bloody nose, which is strange. It must be something to do with the dry air. My second cousin, Jim, rode his bike next to me all day, which is probably why the run

was successful. Every six to ten miles we would take a gas station break and restock on water and PowerAde. At mile 21 we stopped for a nice lunch at a classic New Jersey diner. I ate way too much for still having to run. Jim said I had to try a crab cake since I was on the east coast now. I'm glad I tried one but it didn't sit well for the last six miles."

Between the lunch and the afternoon sun, things went downhill pretty fast. I eventually stopped and sprawled out in the middle of a parking lot. I felt I was literally melting, like a plastic toy on fire. Jim stood over me and said "I'm guessing you're done for the day, I'll call my friend Dave to come and get us."

After a shower, nap and dinner we all went to an outdoor play of Shakespeare's *Hamlet*. It was an epic tale. It was easy to tell when the play was finally over because every single main character lay dead on the stage.

A little over a 100 miles left in my run across America remained. I could barely remember the start; it feels so long ago, a lifetime or more. It was difficult to run well, the heat, traffic and Ruby were not cooperating. We trudged on anyway. I spent the remaining days thinking of the end for motivation, it worked a little, so close yet so far away. My parents, sister Sarah, Josh and Kate had all left Wisconsin yesterday to come out for the end and would be arriving sometime tonight. I had been anticipating the moment all day.

Day 134 - 32 miles
"We stayed in a Motel 8 just before the bridge into Delaware. Having family around and talking about the end of the run really makes it feel like the Atlantic Ocean might actually be out there somewhere."

They had planned to drive down to D.C. and see the sights while I ran. It was another hot one and I was glad to leave Ruby behind. A bank clock read 98 degrees so I was surprised at how well I was running. I have developed a habit in the east of running into McDonalds, buying a small cup for a $1.06 and drinking icy mountain PowerAde until my veins turn blue. I stood in front of the fountain for about five minutes downing cup after cup. Sweat was literally flowing out of my pores, not dripping... flowing. It was not long before a puddle surrounded my feet and the cleaning lady came out with a towel to wipe it up. I moved five feet to the left to let her do her thing. The pattern continued until she said, "Maybe you should just try staying in one spot." I apologized and continued my run.

I was averaging around five miles an hour throughout the day. A summer storm rolled in and the downpour of rain seemed to put the fire out that was burning up my skin. I ran until the police pulled me over and hauled me off the Delaware Memorial Bridge. Apparently to the general public, I looked like someone who was about to commit suicide. Little did they know that I was actually feeling quite good about my current outlook on life. Now if they had seen me at a few of my low points along the route their claims would have probably been more accurate. It was the sixth time I had been pulled over on the run, but it was the first time I had to get in the back of a police car. An old lady was unloading her groceries when the police pulled up next to her to let me out. The look on her face was priceless as he opened the door to let his prisoner out, a half naked barbaric man free to roam the streets once again. She threw the remaining groceries into the trunk without care, hurried to the wheel and drove off without a word. I am sure my cold stare did not help ease her worries but I do tend to get a bit zombie like after a long day of running.

Day 135-136 Atlantic Ocean
"The start of this journey seems like a lifetime ago.

Time has a way of sorting things out and organizing life in a way few people comprehend. During my trek this was a thought I clung to and prayed to be true. I knew if I could somehow keep putting one foot in front of the other, survive just one more day, one more night, eventually time and steps would take me home."

I had been riding an emotional, physical, mental and spiritual roller coaster for months and the end was finally in sight, the end, being the Atlantic Ocean and the conclusion of my wildest and most incomprehensible dream to date. It's hard to describe the feeling of approaching the end of such a legendary journey or the thoughts that went through my head in the closing miles. When I started out with my feet submerged in the Pacific Ocean 136 days ago my mind was filled with questions, wonder and the daunting thought of the unknown. Everyone had the same questions, and the only way I could answer them was to run. As the miles and days passed, questions found their answers, wonder was replaced with memories and the unknown became stories, some of which only I will ever know to be true.

People ask me all the time, "Why did you decide to do the run?" I usually reply by saying, "The adventure aspect intrigued me, and I wanted to do something to help solve the water crisis." Although this is true, it only scratches the surface of my reasoning. Running across America was the biggest dream I had at the time. After fulfilling a smaller dream of fixing up an old sailboat and living on it in Lake Michigan last summer many people started to share their dreams with me. I spoke with so many people who had regrets of dreams unlived. They kept telling me I need to do these things while I'm still young, before life gets a hold of me. Their advice always struck me as odd since they where usually standing alive right in front of me. They were still capable of fulfilling their dreams in my book; however, they usually never saw it that way. If you want to accomplish something, the answer is not tomorrow, not yesterday, it is today. It is always now.

"The place where God calls you is the place where your deep gladness and the worlds deep hunger meet." **Fredrick Buechner**

I have always desired to lead a fulfilled and meaningful life; for some reason my answer to that was to run across America. The philosophy I learned on the sailboat also rang true for the run- follow your dreams and you will not be disappointed. You will not look back and say, "If only I would have lived on a sailboat," or "If only I would have run across America." I know how time works and I know one day I will have to look back. Anything can be accomplished one step at a time. Sometimes the first step is the hardest, but without that first step I would have never been able to take my last step into the Atlantic Ocean.

Above - Letting the moment sink in on the Atlantic City Boardwalk.

Above - Carrying Kate into the Atlantic Ocean against her will.

Below - Cheryl Thornton, my sister Sarah, Kate, Dad, Josh and I

EPILOGUE

The run ended on June 30th 2010 in the predicted 136 days. Abe became the 15th person to run across America solo and unsupported. The run raised over $90,000 for the non-profit Living Water International who exists to demonstrate the love of Jesus Christ in underdeveloped countries who lack access to clean drinking water. "After the massive earthquake that ripped across Haiti in January 2010, Living Water International's work there went into overdrive. The need for clean drinking water was more desperate than ever, with cholera running rampant through overcrowded tent cities and water trucks unable to keep up with demand. After the earthquake, Living Water's output for rehabilitated wells more than doubled (from around 175 to around 450), plus the addition of a new drilling program in Port au Prince and another drill rig headed to Cap Haitian for a new drilling program there." **Living Water International**

Three months later, Abe and Kate were married on the shores of Lake Michigan. Abe worked for a marketing company for five months before deciding a button shirt and tie and being spoon fed the dream of becoming wealthy was not for him. Kate continues to pursue a painting degree from UW-Milwaukee and works as manager of the art center on campus. After a May 2011 mission trip to Haiti was canceled do to political unrest and violence in the area, Abe decided to continue helping raise money for Living Water International through endurance sports. On June 20th 2011 he teamed up with a small group called h2o ride. They began a 9,200 mile bicycling journey that circled the edges of the continental forty-eight. The riders completed the 145 day journey in November 2011 raising over $26,000 for well projects in the drought stricken country of Ethiopia. The mission trip to Haiti has been rescheduled for June of 2012. To request a speaking appearance, follow Abe's current adventure or just drop a note please visit **www.abrahamlouis.com**

Above - Dad and Mom came out to visit me in the plains, here we are standing on the Oklahoma State Line.

Below - Dad and I with a sign that his school made.

Above - Kate helping me pack Ruby before she leaves.

Below - Kate and I sharing the passenger's seat, Josh is driving with the new rabbit and Ruby in the back seat.

Above - At The Crossing, a large church in St. Louis. They annually have over 200 runners that complete the St. Louis GO Half Marathon. They raise thousands of dollars every year for LWI.

Below - I did numerous radio interviews in the towns I passed though, pictured below is at a Christian radio station on the Navajo Indian reservation.

Above - A tour of the Reliv head quarters in St. Louis

Below - Todd and I in Kansas. Todd is one of the funniest people I have ever met, a real comedian and joy to be around. The back of his shirt read "We're all Grumpy in the morning"

Above - In Grafton, WI, a group of students went around selling www.water.cc bracelets after school to raise money for the LWI. Carson (in the under armor shirt) played, myself in a Destination Imagination skit.

Below - The Borchards with their neighbor, Lukas, and I

Above - My body guards through East St. Louis. I stayed at Kerry's house for two nights, he is pictured on the right.

Below - The McCartney Family in Indiana - a few weeks later Pat ran the Western States100 in which he completed in 29 hrs. 16 min.

CPSIA information can be obtained at www.ICGtesting.com
Printed in the USA
LVOW101733120412

277383LV00003B/4/P